Numerical Simulation of Effluent Discharges

Numerical Simulation of Effluent Discharges: Applications with OpenFOAM provides a resource for understanding the effluent discharge mechanisms and the approaches for modeling them. It bridges the gap between academia and industry with a focused approach in CFD modeling and providing practical examples and applications. With a detailed discussion on performing numerical modeling of effluent discharges in various ambient waters and with different discharge configurations, the book covers the application of OpenFOAM in effluent discharge modeling.

Features:

- Discusses effluent discharges into various ambient waters with different discharge configurations.
- Focuses on numerical modeling of effluent discharges.
- Covers the fundamentals in predicting the mixing characteristics of effluents resulting from desalination plants.
- Reviews the past CFD studies on the effluent discharge modeling thoroughly.
- Provides guidance to researchers and engineers on the future steps in modeling of effluent discharges.
- Includes an introduction to OpenFOAM and its application in effluent discharge modeling.

The book will benefit both academics and professional engineers practicing in the area of environmental fluid mechanics and working on the effluent discharge modeling.

IAHR Book

Series editor: Robert Ettema
Department of Civil and Environmental Engineering, Colorado State University, Fort Collins, USA

The International Association for Hydro-Environment Engineering and Research (IAHR), founded in 1935, is a worldwide, independent organisation of engineers and water specialists working in fields related to hydraulics and its practical application. Activities range from river and maritime hydraulics to water resources development and eco-hydraulics, to ice engineering, hydroinformatics and continuing education and training. IAHR stimulates and promotes both research and its application, and, by doing so, strives to contribute to sustainable development, the optimisation of world water resources management and industrial flow processes. IAHR accomplishes its goals by a wide variety of member activities including the establishment of technical committees, working groups, congresses, specialty conferences, workshops and short courses; the commissioning and publication of journals, monographs and edited conference proceedings; involvement in international programmes, such as the UNESCO, WMO, IDNDR, GWP, ICSU and The World Water Forum; and by co-operation with other water-related (inter)national organisations.

www.iahr.org

Energy Dissipation in Hydraulic Structures
Edited By Hubert Chanson

Hydraulics of Levee Overtopping
Lin Li, Farshad Amini, Yi Pan, Saiyu Yuan, and Bora Cetin

Climate Change-Sensitive Water Resources Management
Edited by Ramesh S. V. Teegavarapu, Elpida Kolokytha, and Carlos de Oliveira Galvão

Water Projects and Technologies in Asia
Historical Perspectives
Edited by Hyoseop Woo, Hitoshi Tanaka, Gregory De Costa, and Juan Lu

For more information about this series, please visit: https://www.routledge.com/IAHR-Book/book-series/IAHRMON

Numerical Simulation of Effluent Discharges

Applications with OpenFOAM

Abdolmajid Mohammadian,
Hossein Kheirkhah Gildeh, and Xiaohui Yan

Designed cover image: Shutterstock | grafner.

First edition published 2023
by CRC Press
6000 Broken Sound Parkway NW, Suite 300, Boca Raton, FL 33487-2742

and by CRC Press
4 Park Square, Milton Park, Abingdon, Oxon, OX14 4RN

CRC Press is an imprint of Taylor & Francis Group, LLC

British Library Cataloguing-in-Publication Data
A catalogue record for this book is available from the British Library

ISBN: 978-1-032-02048-8 (hbk)
ISBN: 978-1-032-02094-5 (pbk)
ISBN: 978-1-003-18181-1 (ebk)

DOI: 10.1201/9781003181811

Typeset in Times New Roman MT Std
by KnowledgeWorks Global Ltd.

Contents

Preface

This book is an introduction to the numerical modeling of effluent discharges using OpenFOAM. It introduces the relevant background knowledge and modeling techniques of effluents in detail.

With the increase in population, the development of regional economy and the acceleration of industrialization, wastewater effluents are increasing, which puts forward higher requirements for effluent treatment and disposal capacity. The establishment of a water effluent model can play an important role in improving effluent treatment technologies and dealing with environmental pollution. Therefore, effluent modeling is very important. This book introduces computational fluid dynamics (CFD) models of effluent discharges on the basis of understanding the research progress of jets and plumes, analyzes the advantages and disadvantages of different modeling approaches, and puts forward a series of suggestions for future research work.

Although a great deal of research has been done on the mixing properties of wastewater discharges over the past few decades, the simulation of wastewater discharges using modern mathematical and computational techniques is still in its infancy. The basic data problem and the uncertainty of model parameters in the process of model application pose challenges to the reliability of the model. The availability of open-source CFD tools has opened the door to more realistic CFD modeling of effluent discharges. Although the numerical simulation technology has been significantly developed, the turbulence modeling problem of jet or plume has not been completely studied, and further research is needed. This book discusses these gaps in the literatures.

This book is mainly for the undergraduate and graduate students in hydraulics and hydrology, as well as for practitioners. The book begins with an introduction to outfall systems (Chapter 1) and introduces the reader to the application and different configurations of outfalls, various types of effluents, and mixing zones. Chapter 2 introduces the basic principles of numerical modeling. First, it introduces the governing equations, computational domain, boundary conditions, and initial conditions. Then, computational meshing is introduced, including mesh generation and determination of mesh resolution based on mesh sensitivity analysis. Then various methods in turbulence modeling (RANS, LES, DES, DNS) are briefly discussed. The basic concepts are discussed, and the formulations of the selected methods are provided. Finally, the modification of turbulence terms for buoyant discharges is introduced. Chapter 3 is an introduction to OpenFOAM. OpenFOAM is a popular CFD tool for effluent discharge modeling. It mainly introduces the OpenFOAM solvers and mesh generation and post-processing capabilities for effluent discharge modeling. Chapter 4 reviews past numerical studies in this field, points out future research directions, and puts forward suggestions for further improving effluent discharge modeling.

This book builds on past effluent discharge researches with further discussion on CFD modeling approaches and techniques, and it is hoped that these introductions and recommendations will be a useful reference for undergraduate and graduate students in hydraulics and hydrology, as well as the practitioners.

List of symbols

C	concentration of tracers or contaminants
D	diffusivity coefficient
D	diameter of the discharge port
F_r	the densimetric Froude number
g	gravitational acceleration
g_x, g_y, g_z	gravitational acceleration components in the x, y, and z directions
h_0	water depth (m)
k_{eff}	heat transfer coefficient
p	pressure (N/m^2)
pr_t	turbulent Prandtl number
pr	Prandtl number
S	salinity (psu)
t	time (s)
t_0	initial time (s)
T	temperature
T	fluid temperature (°C)
U_j	initial velocity (m/s)
u	instantaneous velocity component (m/s)
u, v, w	velocity components (m/s)
u_x, u_y, u_z	velocity components (m/s)
v	kinematic viscosity
v_t	turbulent viscosity
x, y, z	coordinates
ω	specific dissipation rate
ρ	density
ρ_α	ambient density (g/cm^3)
ρ_j	jet's initial density
τ_{ij}	stress in the *j*-direction exerted on a plane perpendicular to the *i*-axis (N/m^2)

List of abbreviations

CFD computational fluid dynamics
CORMIX Cornell mixing zone expert system
DES detached eddy simulation
DIC diagonal incomplete Cholesky
DILU diagonal incomplete lower upper
DNS direct numerical simulation
EPA environment Protection Agency
EU European Union
FDIC faster diagonal incomplete-Cholesky
FV finite volume
FVC finite volume calculus
FVM finite volume method
FDM finite difference method
GAMG geometric-algebraic multi-grid
GGDH general gradient diffusion hypothesis
LES large eddy simulation
MSF multistage flash
NS Navier-Stokes equations
OSS open source software
ODE ordinary differential equations
PBiCG preconditioned bioconjugate gradient
PCG preconditioned conjugate gradient
PDE partial differential equation
PDR partial differential equation
PML Prandtl mixing length
RANS Reynolds-averaged Navier-Stokes
RNG renormalization group
RO reverse osmosis
RSM Reynolds Stress Model
RWPT random walk particle tracking
SGDH simple gradient diffusion hypothesis
VTK visualization toolkit
WSL windows subsystem for Linux

Chapter 1

Introduction

1.1 Identifying the problem

Discharges of industrial effluents into coastal and estuarine waters and the emissions of incinerated urban waste into the atmosphere provide two examples of environmental flows in which water and air quality, respectively, are determined by the behavior and structure of the particle-laden, turbulent, dense/buoyant jets, or plumes generated by discharges. Industrial power plants discharge residual byproducts into water bodies (Lattemann and Höpner, 2008), mostly as submerged jets due to their higher effectiveness.

Moreover, rising populations, shortages of clean and potable water, and advancements in desalination plant technology have increased rapidly in recent decades. In arid and semi-arid countries, desalination plants are actively considered as the best alternative to respond to the high demand for drinkable water. Desalination plants remove the dissolved minerals from coastal water bodies and produce effluents with a high-salt concentration, called brines, that may also have an elevated temperature, especially for multistage flash (MSF) desalination plants. Disposal of these brines, which have higher density than the receiving water, causes many environmental impacts, especially in the near field of outfall systems, which is the natural habitat of marine species and fish cultures (Hashim and Hajjaj, 2005; Lattemann and Höpner, 2008). Some areas such as the Red Sea, Persian Gulf, and generally low energy areas with shallow waters are very sensitive to effluent discharges.

The effluent discharge systems of industrial power plants have to be designed properly in order to minimize environmental impacts and financial costs. They also must satisfy the environmental criteria and standards (e.g., US-EPA and EU regulations). Nevertheless, ocean outfall systems are mostly not optimized, either regarding environmental impacts or practical needs. In some cases, regulations also lack clear guidelines for ambient water or effluent standards (Jirka, 2004).

The density differences between the effluent and ambient water, represented by the buoyancy flux, result in various flow and mixing characteristics of the discharge. In the case of dense jets, especially brine from reverse osmosis (RO) desalination plants, the flow has the tendency to fall as negatively buoyant plumes. On the other hand, buoyant jets (e.g., effluents from MSF desalination plants) have lower density than ambient water which causes the plumes to rise.

Besides being designed to minimize environmental impacts and financial costs, discharge outfalls must be in compliance with regulatory criteria. The first step before working on the discharge outfall design is to decrease the concentrations of the waste

DOI: 10.1201/9781003181811-1

source within the industrial plant (e.g., decreasing the additive usage, enhancing plant efficiency, pretreatment technologies, etc.). The second step is the application of improved mixing technologies, like submerged diffusers, and to discharge in less sensitive regions (offshore, deep waters).

Although experimental studies on scaled physical models have been primarily used to study the mixing problems in jets and plumes, this book only focuses on the numerical aspects of such problems. Therefore, the problem at hand is clear: how numerical modeling can be of help to design effluent discharges in open waters more efficiently. This requires an understanding of the problem and knowledge of the tools needed to address the questions surrounding it.

The term "numerical modeling" is still a general term and vague to some extents. There are different numerical modeling techniques that try to solve the equations of transport for effluent discharges. These models may use either simplified or complex sets of partial differential equations (PDEs) such as mass conservation, momentum conservation, and transport equations. There are two main methods available in solving effluent discharge problems numerically that are briefly summarized below.

1.1.1 Integral models

Jet integral models, according to Robinson et al. (2015), solve mass and momentum conservation equations based on the assumptions that the velocity profiles of jets have no radial variation, and that the jet profile is axisymmetric and Gaussian. In other words, complex governing equations of flow hydrodynamics are integrated over the cross section, assuming a Gaussian cross-sectional distribution. These models simplify the PDEs to ordinary differential equations (ODEs) that can be easily solved using numerical integration of differential equations such as explicit and implicit numerical methods. Explicit methods calculate the state of a system at a later time using the state of the system at the current time ($S(t + \Delta t) = F(S(t))$, while implicit methods use both current and future states of the system to find a solution ($F(S(t), S(t + \Delta t)) = 0$). In the 1950s and 1960s, first-order jet integral models were proposed by Morton et al. (1956) and Fan (1967) based on the jet entrainment closure approach and by Abraham (1963) and Turner (1969) based on the jet diffusion approach. Wang and Law (2002), Yannopoulos (2006), and Jirka (2004) developed second-order jet integral models. Since the turbulent mixing of effluent discharges are complex, as are their numerical solutions, integral models rely on experiments to derive the coefficients for their simplified analytical methods.

According to Robinson et al. (2015), the integral models are less reliable when there is any of the following: (i) the discharge's initial momentum and buoyancy acting in opposite directions, resulting in instabilities on the edge, as observed in the inner half (lower half) of inclined dense jets; (ii) noticeable interaction between the mean flow and the jet, (iii) an unsteady mean ambient flow; (iv) a significant effect due to horizontal or lateral boundaries; (v) an unstable near-field area, with a re-entrainment of concentrated effluent into the jet; or, (vi) a large re-entrainment of concentrated effluent from mid- and far-fields into the near-field jet due to tidal cycles.

The most popular integral models in effluent discharge modeling are: CORMIX, VISUAL PLUMES, and VISJET. These models have been reviewed by Palomar et al. (2012) in detail, and the following provides a summary of that study.

CORMIX (Cornell Mixing Zone Expert System) software (Doneker and Jirka, 2001) is a commercial model that was developed in the 1980s at Cornell University (USA) as a project funded by the Environmental Protection Agency (EPA). Supported by the EPA, it became one of the most popular programs for discharge modeling. CORMIX is an expert system for predicting the discharge trajectory and dilution into water bodies in steady-state without considering time series data. CORMIX can simulate the disposal of effluents with positive, negative, and neutral buoyancy under different discharge and ambient conditions. The subsystems CORMIX 1, 2, and 3 are based on dimensional analysis of the processes, while the CORJET model is based on the integration of differential equations.

VISUAL PLUMES by Frick (2004) is a free access software developed by the EPA, which includes several models to simulate positively, negatively, and neutrally buoyant effluents discharged into receiving water bodies. VISUAL PLUMES considers the effluent properties, the discharge configuration, and the ambient conditions (temperature, salinity, and currents whose intensity and direction can be variable through the water column). It is limited to near-field region modeling and does not simulate the interaction of the flow with boundaries. It can consider time series data, simulating discharges in scenarios which change over time.

VISJET (Innovative Modeling and Visualization Technology for Environmental Impact Assessment) software (Cheung et al., 2000) is a commercial model developed by the University of Hong Kong, which can simulate positively and negatively buoyant discharges. VISJET considers the effluent properties, the discharge configuration, and the ambient conditions (temperature, salinity, and currents whose intensity and direction can be variable through the water column). It is limited to near-field region modeling and does not simulate the interaction of the flow with boundaries.

1.1.2 Computational fluid dynamics (CFD) models

Effluent discharge modeling by CFD tools is not perfect, but it is an improvement over the parameter-based jet integral models. Issues that remain with CFD tools include the following: (i) accuracy, (ii) stability, (iii) computational time, (iv) complicated codes that require expert knowledge to use them efficiently and accurately, and (v) simulations that need calibrating and validating.

Turbulent flow models are often resolved with a turbulence model to parameterize unresolved mixing and dispersion scales. One should apply turbulence models with caution, as they sometimes provide stable but unrealistic solutions, such as when they are applied to physical scenarios for which they have not been validated for.

When using a CFD model, it can be a challenge to create and resolve the mesh and to define appropriate boundary conditions (e.g., intensity and turbulence dissipation rate). A high-mesh resolution is often needed for a stable solution, even when the turbulence model is a good match. This means that CFD modeling is computationally expensive. Even with current computing systems, accurate CFD models for near-field dispersion and mixing might need simulation times of several days or weeks. This is much more expensive compared to the integral models that can produce results on the order of minutes and seconds. There is a balance between model stability, numerical diffusion, mass and momentum conservation, boundedness, and computational cost. These choices can significantly influence the estimation of modeled concentration.

However, once built, calibrated, and validated, CFD models can produce high-resolution three-dimensional images of jet mixing and dynamics. CFD models are free from some of the assumptions that restrict integral models. Since CFD models do not require the assumption of a steady-state condition or self-similarity in the jet profile, they can include a variety of external effects such as the presence of surface waves and encompass a wide range of boundary conditions to allow users to directly simulate the boundary interaction.

CFD modeling of jet discharges has been approached in a variety of ways, including both hydrostatic and nonhydrostatic approaches to the Reynolds-averaged Navier-Stokes (RANS) and the Large Eddy Simulations (LES) models. Both models have functioned well over the past decade to simulate effluent discharges. RANS models are based on a time-averaging method and result in a time-averaged mean velocity field, which is averaged over a longer time period than the time constant of the velocity fluctuations, and results in a constant mean velocity without fluctuation for time-dependent variations. LES is based on filtering instead of averaging. A filter size is identified, and flow scales equal to or larger than this size are calculated exactly, and scales smaller than the filter size are modeled. The smaller the filter size, the more concise is the calculated time variation resolution of the velocity vectors. RANS models are more numerically efficient than LES models, while providing enough detail for engineering applications. Thus, they have become the most prevalent CFD models used for the design of outfall systems.

The Direct Numerical Simulation (DNS) method is less applicable to engineering problems, functioning more as a research tool. It is CPU-intensive, as it attempts to resolve Navier-Stokes equations with no approximation of the turbulence and requires a very fine numerical resolution to capture all the turbulence details. It basically resolves entire turbulence scales temporally and spatially. Mesh systems should be very fine to resolve all the spatial scales (Kolmogorov, 1941).

Table 1.1 (after Zhao, Chen and Lee, 2011) summarizes the existing modeling packages (commonly used in the academia and industry) for the simulation of jet and plume mixing.

1.2 Application of outfalls

Outfalls have been used for many years. Initially, they have been used as a means of transporting the effluents to the discharge point, in the absence of environmental regulations. In the modern era, outfalls are used in both inland and coastal waters more carefully, and as a system that increases the dilution of discharged effluents to meet the environmental regulations in both near-field and far-field mixing zones. In other words, outfalls are not simply a method of transport, they represent a sustainable technology to preserve the environment while meeting its main objectives. More restrictive regulations have been developed throughout the years and the design of outfall systems has become more complex.

It is noteworthy that stormwater outfalls that discharge non-impacted waters into river or marine environments are not part of what is discussed here. In this book, we primarily refer to the outfalls that transport and discharge effluents with elevated temperature, salinity, and other chemicals. For instance, outfalls used in desalination plants, nuclear power plants, and wastewater treatment plants are among those mostly studied with respect to mixing problems.

Table 1.1 Existing modeling packages for simulation of jet and plume mixing

Models	*Mathematical approaches for jet/plume mixing*	*Availability*	*Major functionalities and capabilities*
CORMIX	Empirical solutions; Eulerian jet integral method	Commercial package	Prediction of jet and (or) plume geometry and dilution in the near field; single or multiple jets
VISJET	Lagrangian jet integral method	Commercial package	
Visual PLUMES	Empirical solutions; Eulerian and Lagrangian jet integral methods	Free package	
NRFIELD	Empirical solutions	Free package	Prediction of jet and (or) plume geometry and dilution in the near field of multiport diffusers
Sophisticated Multidisciplinary Models			
OpenFOAM	FVM; RWPT method	Free package	Predictions of ocean hydrodynamics; pollutant fate and transport in the near and far fields; water quality; sediment processes
MIKE21/3	FVM; RWPT method	Commercial package	
Delft3D	FDM; RWPT method	Free package	
ANSYS CFX	FVM; RWPT method	Commercial package	
ANSYS Fluent	FVM; RWPT method	Commercial package	
FLOW-3D	FDM; RWPT method	Commercial package	
TELEMAC-2D/3D	FEM; RWPT method	Free package	
EFDC	FDM; RWPT method	Commercial package	Predictions of ocean hydrodynamics; Pollutant dispersion in the far field; Suspended sediment transport
HydroQual–ECOMSED	FDM; RWPT method	Free package	Predictions of ocean hydrodynamics; Pollutant fate and transport in the far field; Sediment processes

Note: FVM: Finite Volume Method, FDM: Finite Difference Method, RWPT: Random Walk Particle Tracking.

Although outfalls are essential to human needs, they impact the environment they discharge into. The direct discharge of wastewater into lakes, rivers, and seas can increase turbidity and change the ambient temperature. Salinity is also a major public and scientific concern. Coastal waters receive concentrated salt brine as discharges from seawater desalination plants, chemical wastes from biofouling (e.g., chlorine), and fertilizers. The water bodies that receive the industrial discharges are often very sensitive environments, and designing outfalls to disperse effluent and reduce the concentration of effluents is essential in helping to protect the receiving water bodies.

Hopner and Windelberg (1996) noted that certain coastal ecological zones are particularly vulnerable to effluent discharges, including salt marshes, mangrove forests, coral reefs, and other low-energy intertidal areas. The Persian Gulf and the Red Sea are particularly sensitive to effluent due to their low hydro-dynamism. Local fisheries, tourism industries, and other economic concerns are affected by the health of coastal environments (Figure 1.1).

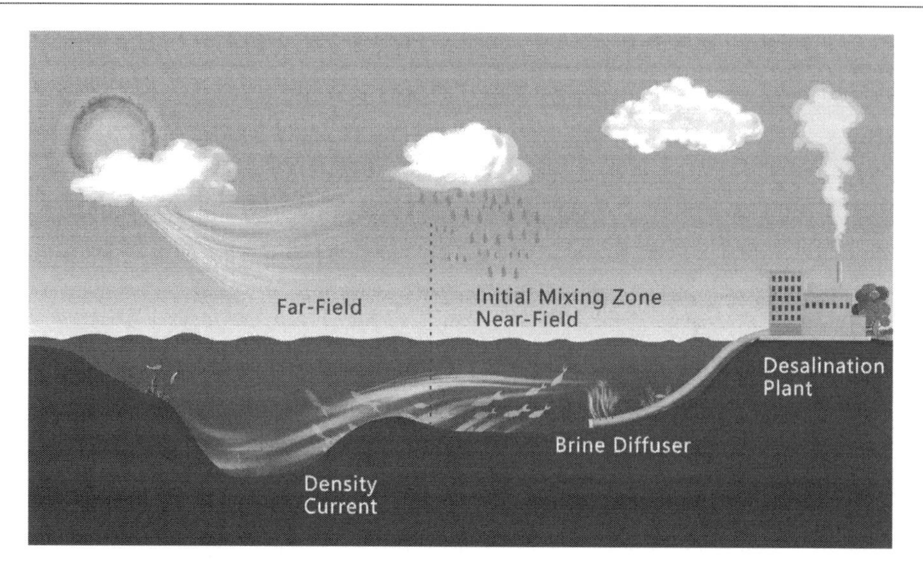

Figure 1.1 An overview of the effluent discharge in a water balance concept.

Based on the above paragraphs, it is clear that outfalls are not only a technical and economic concern. It goes beyond that as it affects the socio-environmental aspects of our lives. It is important that both designers and regulators pay attentions to each single outfall system being designed and constructed, as each outfall has its own challenges and unique characteristics.

A key item to note when reviewing the applications of outfalls for effluent discharges is the discharge objectives that would be the basis for the process leading to the water quality regulations, discharge limits, and design criteria. Figure 1.2 shows the general flowchart of environmental evaluations when designing the proper outfall system for a specific site.

The performance of an ocean or inland effluent discharge outfall is dependent on several factors such as the outfall configuration, topographic and bathymetric conditions of the discharge area, receiving water hydrodynamics, etc. This is the main reason that each outfall should be looked at as a unique design with unique characteristics. The following sections summarize key characteristics of effluent discharges.

1.3 Different outfall configurations

The effluents produced by industrial plants (e.g., brine produced by a desalination plant) can be disposed of in several ways such as discharge in inland open waters, injection into ground wells, discharge into a large evaporation pond, or discharge into coastal waters. There are two general methods of discharging effluents: surface discharge through open channels and submerged discharges through pipes extending into ambient waters. These two methods are illustrated in Figure 1.3. Both these types of discharges intend to increase the dilution and mitigate the environmental impacts.

The selection of the type of outfall is site specific and depends on several parameters such as desalination technology, plant operation and production rate, costs, and environmental considerations. Moreover, the characteristics of discharge (such as density)

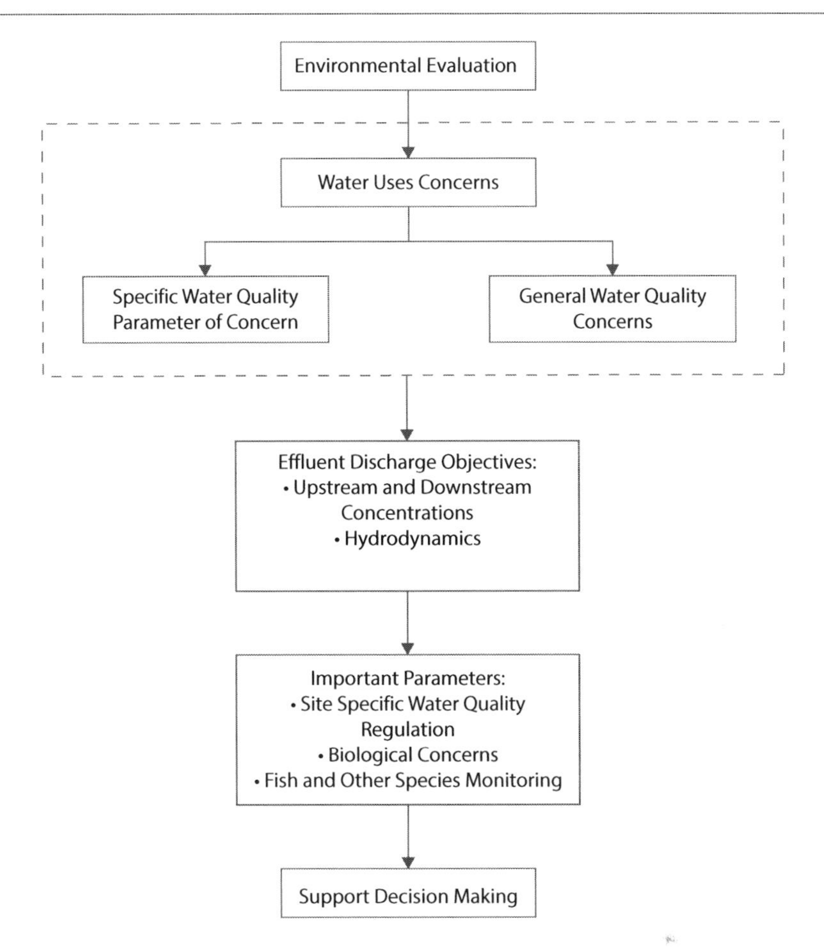

Figure 1.2 Environmental evaluations to preserve aquatic environments from effluent discharge.

and receiving ambient waters (such as density, ambient currents, and buoyancy) influence the type of outfall to be selected.

Surface discharges through open channels are often the most economic options available due to easier construction and maintenance. They are still constructed and used, despite the fact that their efficiency is lower than submerged discharge outfalls in terms of mixing and entrainment (dilution). Surface discharges are excellent options for plants that discharge effluents at a high flow rate, where conveyance through long pipes is extremely difficult due to pipe size and head losses associated with the conduits. Figure 1.4 illustrates the various surface discharge configurations.

On the other hand, submerged discharges are very popular due to their higher efficiency in reaching the required dilution. Therefore, many industrial plants (e.g., mining) will consider the discharge requirements at the very initial stages of the design, and will thus use submerged outfalls in order to meet the regulatory requirements for their permitting purposes. Submerged outfalls often use a pipe near the seabed to discharge effluents. The submerged discharges have been well studied both experimentally and numerically in past decades and our understanding of their mechanisms is relatively well established. It is noteworthy that outfall often refers to the pipe that transports the effluents from

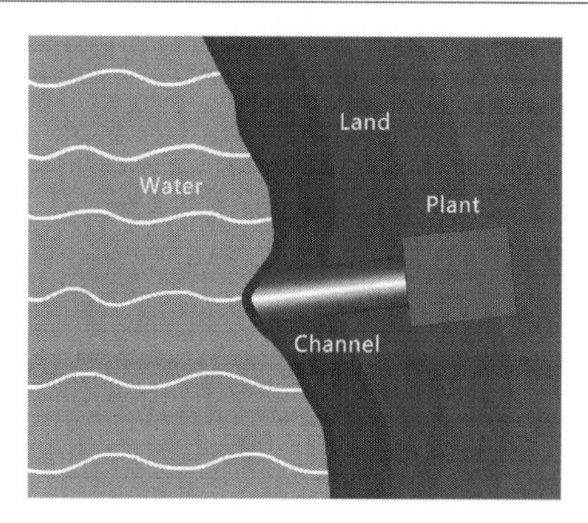

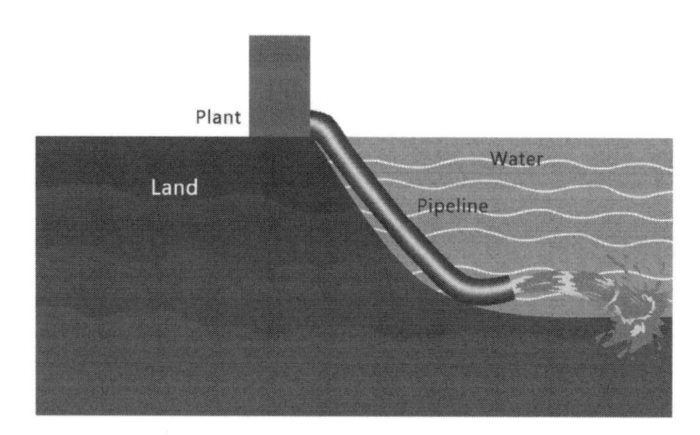

Figure 1.3 Schematic representation of brine discharge systems.

upland into the water. At the discharge location, it can be discharged through the same pipe (i.e., one discharge nozzle) or through a series of nozzles (diffusers). If a single nozzle is used to discharge the effluent into the receiving water, it is called a "single-port diffuser." However, in many cases, the series of nozzles are attached to the discharge point of the outfall, which is called a multiport diffuser (referring to several ports/nozzles to discharge effluents). In this case, the total discharge head

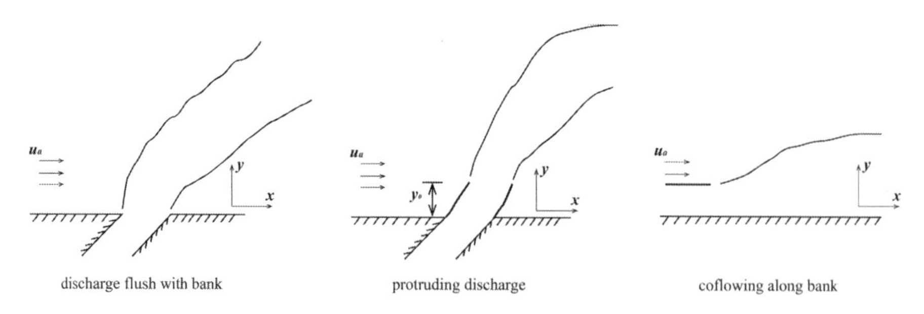

Figure 1.4 Discharge configurations of surface channel relative to bank/shoreline.

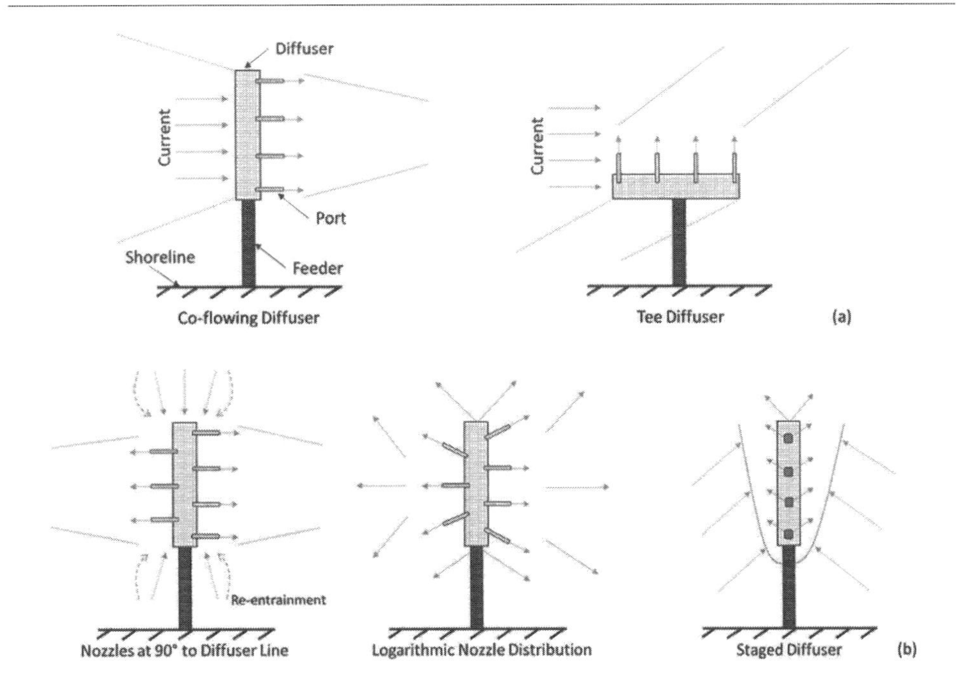

Figure 1.5 Multiport diffusers: (a) Unidirectional diffusers with cross-flow; (b) Alternating diffuser.

is distributed between the number of ports. The configuration of nozzles can vary depending on the ambient condition and design considerations. Figure 1.5 shows five different configurations of multiport diffusers that are commonly used. Another emerging multiport diffuser configuration is the rosette outfall configuration, as shown in Figure 1.6. Unlike the single-port discharges, multiport effluent discharges have been less studied, both experimentally and numerically.

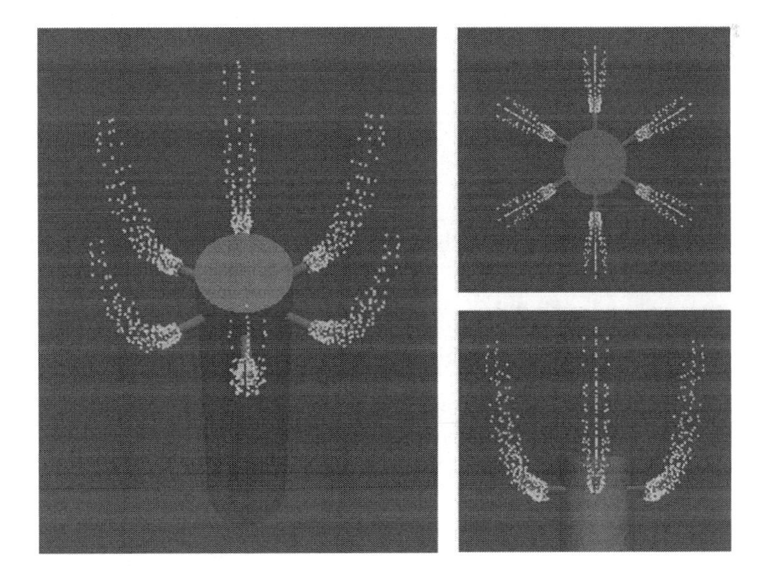

Figure 1.6 Rosette jet, top view.

1.4 Various types of effluents

There are various types of effluents being discharged into the environment. Effluents can be categorized based on the processes that produce them such as municipal/industrial wastewater, combined overflow, drainage water, cooling water, and desalination plant effluent. It is important to realize that the processed effluents from these different facilities will be different in nature and therefore require different and independent considerations.

The other variable in effluent discharges is the physical properties of effluents such as temperature, salinity, viscosity, etc. The discharges from nuclear power plants are often heated effluents while the discharges from oil sands mining are often highly concentrated with little differences in temperature (compared to receiving water).

The effluent flux (volume and momentum flux resulting from flow rate and discharge velocities) is another important variable to consider when designing the outfall systems. This was discussed in previous sections when comparing surface discharges to submerged discharges.

Some effluents may include the chemical/biological properties (virus/bacteria concentrations, etc.) that could result in biological hazards and related concerns. Special pretreatment considerations must be given to such effluents to prevent any hazard to the public.

1.5 Mixing zones

The mixing zone is an important concept in effluent discharge analysis as it has regulatory and legal implications. When discharging into ambient water, the area close to the discharge is expected to be impacted based on chronic exposure to the effluents being discharged. This short zone is often named the mixing zone, which is tolerated environmentally. Water quality criteria can be exceeded in the mixing zone as long as there is no acute toxicity affecting the impacted organisms. At the end of the mixing zone, the water quality requirements set by regulatory agencies should be met and respected. The mixing zone length should be limited as much as possible for the following reasons:

- It should not interfere with the fish spawning zones or shellfish harvesting zones.
- It should not cross the migration path of aquatic organisms.
- It should not attract fish in a way that increases the exposure of organisms to bio-accumulative substances.

To satisfy the abovementioned criteria (just a few of many), mixing zones are defined for effluent discharges based on the physical limits, hydrodynamics of the receiving water, and engineering judgments.

Physical limits are defined in such a way that the system is compliant with the water quality requirements, most of the time. By most of the time, we mean a long duration of time, as the criteria might be exceeded for a short duration of the time due to rare environmental impacts. These limits exist to restrict the length of the mixing zone and to achieve maximum dilution. Some of the common physical mixing limits are summarized below.

- When the mixing process is very slow and the effluent plume spreads over several kilometers, the mixing zone is limited to a maximum length of 300 m such as in slow rivers in Quebec, Canada.

- Unlike above, if the dilution rate is high and effluent mixes with receiving water quickly, the mixing zone is limited to a maximum of 100 m as adopted by several countries such as Peru.

In rivers, the mixing of effluent with the water is relatively rapid. It is, however, important that for river applications, the low flow condition should be considered when accounting for the flow/volume of water in which mixing will occur. Lakes and reservoirs are generally very sensitive to addition of contaminants, as they encourage sedimentation and slow effluent mixing. The estuaries and coastal waters on the other hand are capable of higher mixing rates due to their hydrodynamics and physical variabilities. Regulations of a specific jurisdiction should be reviewed to identify the mixing zone of the discharge location. This is out of the scope of this book. However, there are two main mixing zones based on the general mixing processes and ambient water hydrodynamics, where different physical mechanisms predominate. In the first region, named near-field, the mixing is intense; it results mainly from turbulence generated by the initial buoyancy and momentum of the discharge and the interactions with the ambient flow. Beyond this region, the self-induced turbulence will decay and mixing results only from ambient oceanic turbulence. In this region, which is named far-field, dilution increases at a much slower rate than in the near-field (Figures 1.7 and 1.8). The region in between near-field and far-field is named intermediate field, where buoyant spreading of the plume

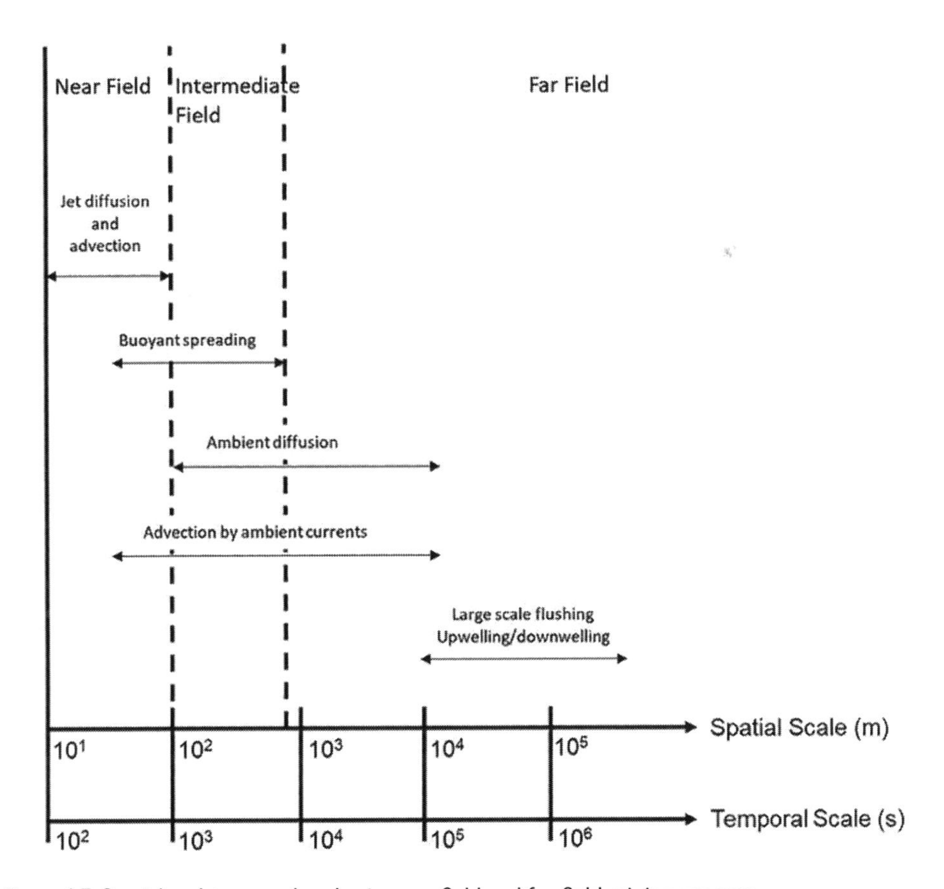

Figure 1.7 Spatial and temporal scales in near-field and far-field mixing process.

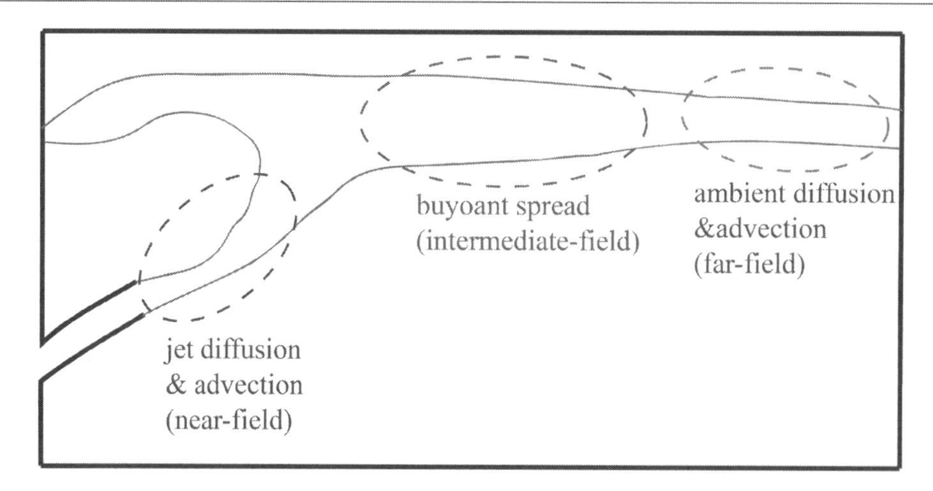

Figure 1.8 Three major domains in positively buoyant jet of a submerged discharge.

happens. While touching upon the intermediate and far-field of plume dispersion, this book mainly discusses the modeling techniques for the near-field region as the most important mixing region due to larger environmental impacts.

1.6 Scope of the book

This book is intended to serve as a resource to university students (undergraduate and graduate), engineers, and scientists interested in or working on numerical modeling of effluent discharges. This book is presented in four chapters as detailed below.

Chapter 1 is a concise introduction of the problem at hand and the background of physical phenomena that are being studied numerically. This chapter will introduce the applications of outfalls and discusses the different common configurations for outfalls in inland and coastal waters. Various types of effluents will also be reviewed in this chapter. It is a challenging task to characterize all effluent types that might exist, both natural and processed, but some fluid properties such as temperature, salinity, and chemical properties will be discussed to categorize the effluents to help with effluent discharge modeling. Finally, mixing zones of the effluent discharges in ambient waters will be discussed. A brief description of each chapter of this book is also discussed in Chapter 1.

Chapters 2 and 3 are the backbone of the book. These chapters introduce the numerical details of simulating effluent discharges numerically. Chapter 2 is an introduction to numerical modeling describing the governing equations, model domain, and choices of boundary conditions. Grid generation for solving the equations and their sensitivity is also discussed in Chapter 2. It also summarizes different turbulence closures. The turbulence models play an important role in the performance of numerical models, and their effects have been widely studied in the literature over the past decade. RANS, LES, Detached Eddy Simulation (DES) and DNS are among the turbulence models discussed in Chapter 2. Modifications of turbulence terms for buoyant discharges are also discussed in Chapter 2.

Chapter 3, as the book's title suggests, is an introduction to OpenFOAM, as the most popular CFD tool in effluent discharge modeling. A decade ago, OpenFOAM was only an academic tool for studying complex fluid mechanics problems. However, it has grown quickly in recent years and has found its way into industry as well (e.g., many consulting firms have invested in creating their own solvers for the particular problems they often solve). This book covers the fundamentals of OpenFOAM related to effluent discharge modeling: the choice of available solvers and differences between them, mesh generation options, and methodology in OpenFOAM and postprocessing the numerical results.

Chapter 4 reviews the application of simulating effluent discharges in water bodies. Past numerical studies will be reviewed and the path forward for filling the research gaps is identified. Chapter 4 also provides some conclusions and recommendations outlined in this book that can hopefully help students and practitioners working on effluent discharge problems, and help to address one of the challenges that human beings face: how to preserve our resources with limited impacts.

References

Abraham, G. Jet Diffusion in Stagnant Ambient Fluid. Ph.D. Thesis, TU Delft, Delft, The Netherlands, 1963.

Cheung, S.K.B., Leung, D.Y.L., Wang, W., Lee, J.H.W., and Cheung, V. VISJET – A computer ocean outfall modeling system. *Proceedings of the International Conference on Computer Graphics*, IEEE Computer Society, 2000.

Doneker, R.L. and Jirka, G.H. CORMIX-GI systems for mixing zone analysis of brine wastewater disposal. *Desalination* 2001, 139, 263–274.

Fan, L.N. Turbulent Buoyant Jets into Stratified or Flowing Ambient Fluids. Ph.D. Thesis, California Institute of Technology, Pasadena, CA, 1967.

Frick, W.E. Visual Plumes mixing zone modeling software. *Environ. Model. Softw.* 2004, 19, 645–654. http://www.epa.gov/ceampubl/swater/vplume/.

Hashim, A. and Hajjaj, M. Impact of desalination plants fluid effluents on the integrity of seawater, with the Arabian Gulf in perspective. *Desalination* 2005, 182, 373–393.

Hopner, T. and Windelberg, J. Elements of environmental impact studies on costal desalination plants. *Desalination* 1996, 108, 11–18.

Jirka, G.H. Integral model for turbulent buoyant jets in unbounded stratified flows, Part 1: The single round jet. *Environ. Fluid Mech.* 2004, 4(2004), 1–56.

Kolmogorov, A.N. The local structure of turbulence in incompressible viscous fluid for very large Reynolds number (in Russian). *Cr. Acad. Sci. URSS* 1941, 30, 301–305.

Lattemann, S. and Höpner, T. Environmental impact and impact assessment of seawater desalination. *J. Desal.* 2008, 220, 1–15.

Morton, B.R., Taylor, G.I., and Turner, J.S. Turbulent gravitational convection from maintained and instantaneous sources. *Proc. R. Soc. Lond. Ser. A Math. Phys. Sci.* 1956, 234, 1–23.

Palomar, P., Lara, J.L., Losada, I.J., Rodrigo, M., and Alvarez, A. Near-field brine discharge modeling part 1: Analysis of combined tools. *J. Desalination* 2012, 290, 14–27.

Robinson, D., Wood, M., Piggott, M., and Gorman, G. CFD modeling of marine discharge mixing and dispersion. *J. Appl. Water Eng. Res.* 2015, 4, 152–162.

Turner, J.S. Buoyant plumes and thermals. *Annu. Rev. Fluid Mech.* 1969, 1, 29–44.

Wang, H. and Law, A.W.-K. Second-order integral model for a round turbulent buoyant jet. *J. Fluid Mech.* 2002, 459, 397–428.

Yannopoulos, P.C. An improved integral model for plane and round turbulent buoyant jets. *J. Fluid Mech.* 2006, 547, 267–296.

Zhao, L., Chen, Z., and Lee, K. Modeling the dispersion of wastewater discharges from offshore outfalls: A review. *Environ. Rev.* 2011, 19, 107–120.

Chapter 2

An introduction to numerical modeling

2.1 Governing Equations

Fluid motion can be described using equations for mass and momentum conservation. These equations can be written in different forms, and can be classified as one-dimensional (1D), two-dimensional (2D), and three-dimensional (3D) equations. Most cases of effluent discharges exhibit obvious 3D features, and thus 3D governing equations are the most popular for simulating efflent discharges. A form of the 3D Reynolds-averaged Navier-Stokes (RANS) equations for mass and momentum conservation can be expressed as follows.

Continuity:

$$\frac{\partial}{\partial x}(u_x)+\frac{\partial}{\partial y}(u_y)+\frac{\partial}{\partial z}(u_z)=0 \tag{2.1}$$

Momentum in the x-direction:

$$\begin{aligned}&\frac{\partial(\rho u_x)}{\partial t}+\frac{\partial}{\partial x}(\rho u_x u_x)+\frac{\partial}{\partial y}(\rho u_y u_x)+\frac{\partial}{\partial z}(\rho u_z u_x)\\&=-\frac{\partial p}{\partial x}+\frac{\partial \tau_{xx}}{\partial x}+\frac{\partial \tau_{yx}}{\partial y}+\frac{\partial \tau_{zx}}{\partial z}+\rho g_x\end{aligned} \tag{2.2}$$

Momentum in the y-direction:

$$\begin{aligned}&\frac{\partial(\rho u_y)}{\partial t}+\frac{\partial}{\partial x}(\rho u_x u_y)+\frac{\partial}{\partial y}(\rho u_y u_y)+\frac{\partial}{\partial z}(\rho u_z u_y)\\&=-\frac{\partial p}{\partial y}+\frac{\partial \tau_{xy}}{\partial x}+\frac{\partial \tau_{yy}}{\partial y}+\frac{\partial \tau_{zy}}{\partial z}+\rho g_y\end{aligned} \tag{2.3}$$

Momentum in the z-direction:

$$\begin{aligned}&\frac{\partial(\rho u_z)}{\partial t}+\frac{\partial}{\partial x}(\rho u_x u_z)+\frac{\partial}{\partial y}(\rho u_y u_z)+\frac{\partial}{\partial z}(\rho u_z u_z)\\&=-\frac{\partial p}{\partial z}+\frac{\partial \tau_{xz}}{\partial x}+\frac{\partial \tau_{yz}}{\partial y}+\frac{\partial \tau_{zz}}{\partial z}+\rho g_z\end{aligned} \tag{2.4}$$

DOI: 10.1201/9781003181811-2

where ρ is the density; t is the time; u_x, u_y, and u_z are the velocity components in the x, y, and z directions, respectively; p is the pressure; τ_{ij} denotes a stress in the j-direction exerted on a plane perpendicular to the i-axis; and g_x, g_y, and g_z are the gravitational acceleration components in the x, y, and z directions, respectively. The turbulence closures for these equations will be introduced in Chapter 4.

2.2 Model domain, boundaries, and initial conditions

2.2.1 Model domain

The model domain is the area to be considered by the numerical models. The fluid model domain can be classified into two types depending on whether the cases of interest focus on external flows or internal flows. For external flows, the computational area is a domain that can sufficiently cover the geometrical object. For internal flows, the computational area is simply the confines of the geometry of the object. It is interesting that the computational domains for problems of effluent discharges can be configured using either approach. For example, in simulations of jets or plumes in laboratory water tanks, the case can be regarded as an internal-flow case, and the computational domain can be simply configured to be consistent with the geometry of the water tank. However, for the near-field simulation of effluent discharges in unbounded ambient conditions, the case can be considered as an external-flow case, and the computational area can be set up as a domain that can sufficiently cover the diffuser and the area of interest.

For simulations performed using OpenFOAM, the computational domain can be configured using either its mesh generation utility, blockMesh, or some other utilities, such as snappyHexMesh and Salome. For simple cases, blockMesh is the most efficient tool for defining model domains. For example, in simulations of jets in cubic water tanks, the model domain can be defined by blockMesh as a cube with six patches, i.e., a bottom patch corresponding to the bottom of the tank, the front, back, left, and right patches modeling the sides of the water tank, and an upper patch representing the water-air interface or the top of the tank. For complicated cases, the computational domain should be created using some professional tools with some special operations, such as cut, rotate, and Boolean.

2.2.2 The boundaries

For a general unsteady problem, the solution conditions include boundary conditions and initial conditions. Boundary conditions are the conditions that the governing equations should satisfy at the fluid motion boundary, and generally have an important impact on numerical calculations. Basic principles for setting common boundary conditions include the following:

1 Setting boundaries at import and export can be beneficial to convergence.
2 There should be no large parameter gradient perpendicular to the boundary.

3 The mesh distortion near the boundary must be reduced, otherwise the calculation error will be too large.

Theoretically, there are three boundary conditions used in OpenFOAM as described below:

1 Assign the value of a variable τ, such as the temperature of the wall, so that the velocity component of the nonsliding wall is zero, etc.
2 Give the derivative value of τ along a certain direction, such as the heat flux of the known wall.
3 The relationship between τ and heat and mass flow is given such that the heat transfer at the wall is limited by the surface heat transfer coefficient and the temperature of the surrounding fluid.

In CFD simulations, common boundary types include the following:

1 Inlet boundary conditions
These are used to specify the value of the flow variable at the inlet. Common inlet boundary conditions include velocity, pressure, temperature, and mass flow inlet boundary conditions.
2 Outlet boundary conditions
Pressure outlet boundary condition is used to define the static pressure at the flow outlet (and other scalars in the return flow). When backflow occurs, using the pressure outlet boundary condition instead of the mass outlet condition often results in a better rate of convergence.
Mass outlet boundary condition is used to simulate the flow when the velocity and pressure of the flow outlet are unknown until the flow problem is solved.
3 Solid wall boundary conditions
For the viscous flow problem, the wall can be set as the no-slip boundary condition, or the tangential velocity component of the wall can be specified (when one face is in translation or rotation), and the wall shear stress can be given to simulate the wall slip. The wall shear stress and heat transfer with the fluid can be calculated according to the local flow conditions. Wall thermal boundary conditions include fixed heat flux, fixed temperature, heat transfer coefficient to external radiation heat transfer, convective heat transfer, etc.
4 Symmetrical boundary conditions
Symmetrical boundary conditions apply when the physical region of the calculation is symmetric. On the symmetry axis or symmetry plane, there is convective flux, and the velocity component perpendicular to the symmetry axis or symmetry plane is zero. Therefore, on a symmetric boundary, the velocity component of the vertical boundary is zero, and the gradient of any quantity is zero.
5 Periodic boundary conditions
Periodic boundary conditions can be used if the geometric boundaries of the flow, salinity, and heat transfer are periodically repeated.

2.2.3 Initial conditions

Initial conditions are used to calculate the initial given parameters, i.e., when $t = t_0$, the function distribution of each unknown quantity is given, such as:

$$\begin{cases} u = u(x, y, z, t_0) = u_0(x, y, z) \\ v = v(x, y, z, t_0) = v_o(x, y, z) \\ w = w(x, y, z, t_0) = w_o(x, y, z) \\ p = p(x, y, z, t_0) = p_o(x, y, z) \\ \rho = \rho(x, y, z, t_0) = \rho_o(x, y, z) \\ S = S(x, y, z, t_0) = S_o(x, y, z) \\ T = T(x, y, z, t_0) = T_o(x, y, z) \end{cases} \quad (2.5)$$

where t is the time; t_0 is the initial time; u is the instantaneous velocity component; v is the kinematic viscosity; ω is the specific dissipation rate; ρ is the density; p is the pressure; T is the temperature; and S is salinity. Subscript 0 is each state quantity corresponding to the initial time.

The unsteady state problem is regarded as being in some stable state at the initial moment. The initial condition of an unsteady problem can be a constant value or a function of spatial location. When the fluid motion is constant, there is no initial condition problem.

2.3 Grid generation and sensitivity analysis

2.3.1 Grid generation

Grids are divided into two categories: structured grid and unstructured grid. Structural grid means that the nodes in the grid are arranged in an orderly manner, the grid points in a certain area that can be numbered uniformly, and the positional relationship between them can be expressed. In an unstructured grid, the positions of nodes cannot be named in an orderly manner with a fixed rule.

Structural grids are simple and regular geometric models. They save storage space, are beneficial to programming calculations, and are easy to handle, but have poor adaptability to complex flow fields. Unlike structured grids, unstructured grids have excellent adaptability, but the generation process is more complicated, especially for flow field calculation problems with complex boundaries.

Regardless of whether it is a structured mesh or an unstructured mesh, the following process is required to generate the mesh:

1. Build a geometric model. The geometric model is the vector of meshes and boundaries. For 2D problems, the geometric model is a 2D surface; for a 3D problem, the geometric model is a 3D solid.
2. Divide the grid. The mesh is obtained by dividing the surface or volume by applying a specific mesh type, mesh element, and mesh density to the generated geometric model.
3. Specify the name and type of the boundary faces.

The key to generating a mesh is to divide the mesh. Since traditional CFD is based on structured grids, there are many mature generation techniques for structured grids, while the generation techniques for unstructured grids are more complex.

2.3.2 *Grid sensitivity analysis*

Increasing the grid resolution within a certain range can improve the calculation accuracy, but increasing the calculation accuracy beyond a certain limit will require a larger computational cost; reducing the grid resolution within a certain range can improve the calculation speed, but beyond a certain limit, decreasing the resolution after the limited value will greatly affect the calculation accuracy. The purpose of the mesh dependence analysis is to identify whether the CFD calculation results depend on the mesh quality, and to seek a balance between calculation accuracy and calculation cost to ensure the correctness of the CFD model. It is an important step in the performance and efficiency of the grid, especially before the grid is used in large-scale engineering calculations, the grid dependence analysis should be carried out to make the selected grid both computationally accurate and economical.

2.4 Rigid lid and free surface boundaries

2.4.1 *Rigid lid*

The rigid-lid approximation is a common simplification in oceanographic studies of medium-density stratified fluids. The so-called "rigid lid approximation" assumes that the free liquid surface is a movable rigid cover, and its boundary conditions satisfy the nonpenetrating condition. It is only suitable for the case in which the free surface fluctuation is not large. If the local change of the free surface water flow and the water flow with a large free surface fluctuation range are studied, the rigid lid approximation can no longer be applied, so another method to determine the pressure change is used: the "elastic cover method" of the free surface; however, the elastic cover method also has its limitations.

2.4.2 *Free surface boundaries*

The free surface is a special interface with its own inherent properties. On the one hand, it is the boundary of the fluid surface, which is a necessary condition for the solution of the flow field; on the other hand, its position is not predicted in advance, but is part of the problem solution, given by the solution process. The free surface is the most basic and extensive research field in hydrodynamics, and its research has profound theoretical and practical engineering significance.

Free surface flow is a common phenomenon in nature. To study the free surface problem by numerical simulation method, it is necessary to deal with two problems: (i) the dispersion of the free boundary and (ii) the movement and development of the free boundary with time. Due to the existence of motion boundaries, the boundary position is only determined at the initial moment, and then the boundary position needs to be solved as part of the calculation results. Therefore, the numerical simulation of free surface flow has always been a major problem in the field of computational fluid dynamics, and it generally requires special processing. At present, the methods used to simulate free surfaces mainly include the elevation function method, the MAC method, and the VOF method.

1 Elevation function method
   If the water depth is a function of space coordinates (x, y) and time t, this function is called an elevation function. The free boundary must move with the fluid and

satisfy the corresponding motion control equation. By solving the water depth along the water flow, the position of the free water surface can be obtained. This method can be used to solve the problem of nonconstant free water surface, but it is incapable of solving the problems of multivalued functions (such as drainage and jet) where the free surface is a coordinate.

2 MAC method
The MAC method was proposed by Harlow and Welch (1965). The main idea is to imagine that there are small particles with no volume and no mass distributed in the flow field, called marking points. These marking points follow the water flow and move with the fluid at the speed of the local flow field. The outer line (surface) of the marked point is the position of the free surface. The MAC method determines the movement position of the fluid free boundary by determining the spatial positions of these marked points at different times. The outstanding advantage of MAC is that it can deal with the problem that the free surface is a multivalued function of coordinates and can vividly describe the flow state evolution of water flow with a free surface, which is of great significance for finely simulating some complex water flow phenomena common in fluid mechanics. The main disadvantage of MAC is that the time-varying coordinate values of all markers must be stored, which greatly increases the computational storage.

3 VOF method
The VOF method was proposed by Hirt and Nichols (1975). The basic idea of the method is to define functions a and b to represent the volume fraction (relative proportion of volume) of water and gas in the calculation area, respectively. In each cell, the sum of the volume fractions of water and gas is 1, that is: $a + b = 1$.
For a computing unit, there are three following situations:

$a = 1$ means the unit is completely filled with water,
$a = 0$ means the cell is completely filled with gas, and
$0 < a < 1$ means that the unit is partly water, partly air, and has a water-air interface

The gradient of the volume fraction of water can be used to determine the normal direction of the free boundary. After calculating the value of each element and its gradient, the approximate position of the free boundary in each element can be determined. Compared with the previous two methods for determining the free surface, the advantage of the VOF method is that it can describe various complex changes in the free surface with only one function. This method not only has the advantages of the MAC method but also overcomes the computational memory problem inherent in the MAC method. It also overcomes the disadvantage that the elevation function method cannot deal with the fact that the free surface is a coordinate multivalued function.

2.5 Introduction to turbulence modeling

Turbulent flows are the most commonly observed natural flows, which can be seen in many situations like outfall systems, recirculating flows, rivers, density currents in dams, estuaries and lakes, atmospheric and oceanic flows, etc. Turbulent flows are highly transient and contain eddies from very small Kolmogorov scales to very large scales comparable with the flow domain.

Turbulence can be studied using different approaches, e.g., field, experimental, analytical, and numerical methods. Turbulence is an inherently 3D phenomenon with a transient and nonlinear nature. This feature of turbulence causes the analytical solutions for turbulent flows to be limited to simplified problems. When subjected to natural and complex fields, the governing equations have no analytical solutions and require numerical computations. On the other hand, the experimental studies of turbulence are limited to specific goals and objectives, and the results are highly dependent on the experimental conditions and vary from case to case. In the meantime, numerical studies of turbulent flows have been developed and have progressed extensively due to the rapid advancement of computational resources. In a review of turbulence (Rodi 2017), it was highlighted that computational speeds for massive parallel computing have progressed considerably over the past 60 years. This has a direct impact on the accuracy of numerical methods for turbulence modeling. This chapter presents a brief overview of various approaches in turbulence modeling.

In many engineering applications, the information about the detailed turbulence fluctuations is not important, and the averaged scales of the turbulence could also be sufficient. Therefore, some scales of turbulence can be neglected. This assumption results in the reduction of the size and nonlinear features of the equations. However, the impact of nonresolved eddies should be modeled and entered into the flow equations using the turbulence models. The more universal, affordable, robust, and accurate the turbulence models are, the higher their applications in the different disciplines of computational fluid dynamics.

2.6 Direct numerical simulation (DNS)

Turbulence modeling is possible mainly using numerical solutions to Navier-Stokes (NS) equations, as they represent all features of complex turbulent motions. Resolving these eddies' entire temporal and spatial scales needs direct numerical simulation (DNS) of NS equations. In this approach, the computational grid should be extremely fine to capture any scale (time, length, and velocity) of turbulence.

The required operations to perform DNS are approximately proportional to Re^3 (Davidson 2015), where *Re* is the Reynolds number. This means that for engineering applications, DNS of turbulent flows with high *Re* numbers is impossible because resolving the entire time and length scales of turbulence needs a massive level of computation. Nelson and Fringer (2018) performed DNS to simulate the wave and current-driven sediment transport in a flume of 1.5 m × 0.312 m × 0.1 m, using about 32 million grid points. They mentioned that using a 480-core supercomputer, simulation of each wave period (3 seconds) took approximately 25 wall clock minutes or 200 CPU hours. Currently, DNS is an applicable method for flows with low *Re* numbers, but it is expected that in the next 20 years, DNS will also become a practical approach for simulation of flows with high *Re* numbers (Rodi 2017).

2.7 Reynolds-averaged Navier-Stokes (RANS) models

A practical approach for the study of turbulent flows is time-averaging of NS equations, called RANS equations. In this approach, it is assumed that the time-dependent variations of turbulence are not important, and the mean impact of turbulence in the

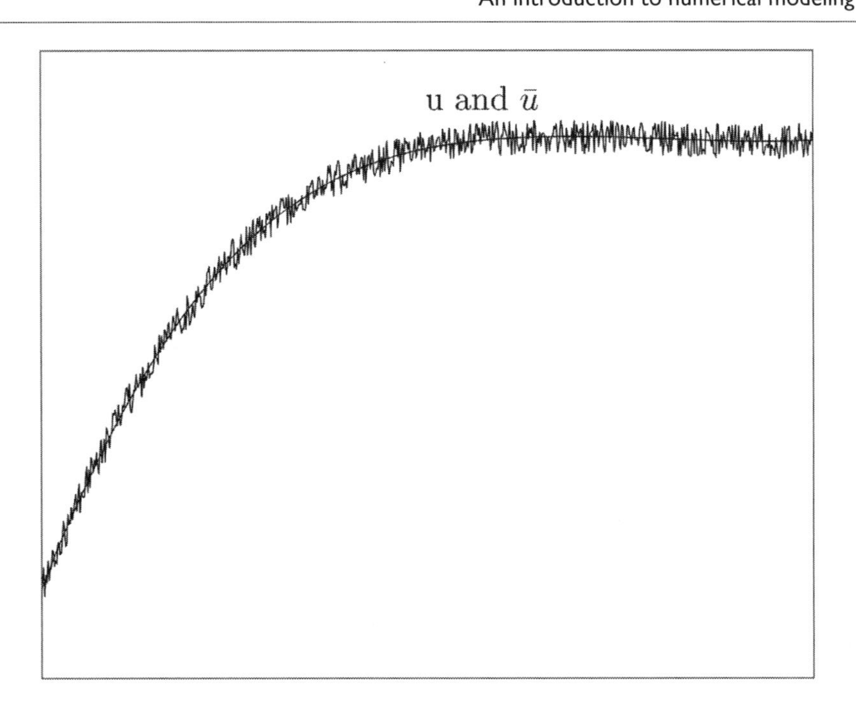

Figure 2.1 Time-averaging of turbulent signal (u) and substitution with time-averaged signal.

flow field is the main factor. This approximation leads to a significant simplification of the NS equations, and is an acceptable approach for engineering applications.

In time-averaging, the velocity (and pressure and density) components are decomposed into mean and fluctuation parts. For example, for the velocity, the decomposition procedure leads to

$$u = \overline{u_i} + u_i' \tag{2.6}$$

where u is the velocity component, $\overline{u_i}$ is the mean part, and u_i' is the fluctuation part (Figure 2.1). However, due to the nonlinearity of NS equations, the process of splitting up and integration leads to the emergence of new unknown terms in the RANS equations, namely Reynolds or turbulent stresses $(\overline{u_i' u_j'})$. These additional terms are different than viscous physical stresses and should be solved using statistical methods. They could be calculated directly via stress transport models or be modeled by the eddy viscosity concept. In the latter approach, the impact of all turbulent motions is added to the systems of equations as an additional viscosity, which is called eddy viscosity Therefore, the RANS equation can be expressed as:

$$\nabla \cdot \boldsymbol{U} = 0 \tag{2.7}$$

$$\frac{\partial \boldsymbol{U}}{\partial t} + \nabla \cdot (\boldsymbol{U}\boldsymbol{U}) = -\frac{1}{\rho_0}\nabla p + (\nu + \nu_t)\nabla^2 \boldsymbol{U} + \frac{\rho \boldsymbol{g}}{\rho_0} \tag{2.8}$$

In other words, the eddy-viscosity-based RANS equations are like the NS equations, except that the impact of turbulent motion is added via turbulent viscosity ν_t, which is computed by turbulent closures.

Extensive research has been performed on the calculation of eddy viscosity, and several models have been proposed in the literature. A large family of turbulence models calculates the eddy viscosity by a length scale and velocity scale. They are classified as zero-equation models, one-equation models, two-equation models, and multiequation models (Rodi 1993). Some other Reynolds models are also proposed in the literature (e.g., Reynolds Stress Models, RSM).

Early RANS models calculate eddy viscosity using constant or algebraic expressions. One of the pioneering works in this regard is the Prandtl Mixing Length (PML) model, proposed by Prandtl (1925), which has been widely used in various areas of fluid mechanics. The PML model is a constant eddy model classified as a zero-equation model, as no PDE is solved to estimate eddy viscosity. This model is robust, simple, and still in use for considering the impact of turbulent dissipation in general applications. It prescribes velocity and length scale as algebraic expressions. However, the model does not consider the transport and history of the turbulence. In the PML, the estimation of mixing length has an empirical basis, making its application difficult for the complex velocity fields. Another fundamental algebraic model is also proposed by Elder (1959) for shallow water flows. It is a simple eddy viscosity model based on bed roughness. This model is embedded in many engineering tools. However, Elder's model is based on bed roughness and does not consider the conditions of flow.

In one-equation models, the velocity scale is determined by solving an advection-diffusion equation (a PDE) for the turbulent kinetic energy (k). This preserves the history and transport of the turbulence in the system. A popular model that is also used in some models is the K model, which is also called the Prandtl-Kolmogorov model (Kolmogorov 1941). The length scale is still calculated using the mixing length theory in this model. Thus, in a sense, it is a semi-empirical turbulence model for the calculation of eddy viscosity.

In two-equation models, PDEs are solved for both velocity and length scales. The k-ε model (Launder and Spalding, 1974) is the most common two-equation RANS turbulence model. It is implemented in many comprehensive CFD models like OpenFOAM and ANSYS-Fluent and comprehensive engineering modeling systems like Delft3D-FLOW (Deltares 2014), Telemac-3D (Hervouet 1999), ROMS-3D (Warner et al. 2008), and Mike3 FM (DHI 2017). In the following, some of the most popular two-equation RANS models are introduced and discussed.

2.7.1 The standard k-ε model

The k-ε model is the first modern turbulence model, introduced by Launder and Spalding (1974). It is still used extensively in the literature and presents reasonable results for many applications. In the k-ε model, velocity and length scales are determined by solving two PDEs for the production k and dissipation rate ε of the turbulent kinetic energy of the turbulence where the eddy viscosity is calculated as:

$$\nu_t = c_\mu \frac{k^2}{\varepsilon} \tag{2.9}$$

where C_μ is a constant coefficient (0.09), and k and ε are determined from the following transport equations:

$$\frac{\partial k}{\partial t} + \frac{\partial k u_i}{\partial x_i} - \frac{\partial}{\partial x_i}\left(D_{keff} \frac{\partial k}{\partial x_i} \right) = G - \varepsilon \tag{2.10}$$

$$\frac{\partial \varepsilon}{\partial t}+\frac{\partial \varepsilon u_i}{\partial x_i}-\frac{\partial}{\partial x_i}\left(D_{\varepsilon eff}\frac{\partial \varepsilon}{\partial x_i}\right)=c_{1\varepsilon}\frac{\varepsilon}{k}G-c_{2\varepsilon}\frac{\varepsilon^2}{k} \tag{2.11}$$

$$D_{keff}=v_t+v \tag{2.12}$$

$$D_{\varepsilon eff}=\frac{v_t}{\sigma_\varepsilon}+v \tag{2.13}$$

$$G=2v_tS_{ij}S_{ij} \tag{2.14}$$

$$S_{ij}=\frac{1}{2}\left(\frac{\partial u_j}{\partial x_i}+\frac{\partial u_i}{\partial x_j}\right) \tag{2.15}$$

where σ_ε, $c_{1\varepsilon}$, $c_{2\varepsilon}$, and c_μ are model constants equal to 1.3, 1.44, 1.92, and 0.09, respectively; k is the turbulent kinetic energy; ε is the turbulent energy dissipation rate; u_i is the time-averaged velocity component in the direction x_i; D_{keff} and $D_{\varepsilon eff}$ are the effective diffusivity values for k and ε, respectively; v_t is the turbulent kinematic viscosity; v is the kinematic viscosity; G is the production of turbulence due to shear; and S_{ij} is the strain-rate tensor. The reader is referred to Rodi et al. 1993) for further details.

2.7.2 The RNG k-ε model

In the RNG k-ε model, the v_t is calculated the same as the standard k-ε model, but the coefficients are calculated differently:

$$\frac{\partial k}{\partial t}+\frac{\partial k u_i}{\partial x_i}-\frac{\partial}{\partial x_i}\left(D_{keff}\frac{\partial k}{\partial x_i}\right)=G-\varepsilon \tag{2.16}$$

$$\frac{\partial \varepsilon}{\partial t}+\frac{\partial \varepsilon u_i}{\partial x_i}-\frac{\partial}{\partial x_i}\left(D_{\varepsilon eff}\frac{\partial \varepsilon}{\partial x_i}\right)=(c_{1\varepsilon}-R_\varepsilon)\frac{\varepsilon}{k}G-c_{2\varepsilon}\frac{\varepsilon^2}{k} \tag{2.17}$$

$$D_{keff}=\frac{v_t}{\sigma_k}+v \tag{2.18}$$

$$R_\varepsilon=\frac{\eta\left(1-\frac{\eta}{\eta_0}\right)}{1+\beta\eta^3} \tag{2.19}$$

$$\eta=\sqrt{S_2}\,\frac{k}{\varepsilon} \tag{2.20}$$

where σ_k, σ_ε, $c_{1\varepsilon}$, $c_{2\varepsilon}$, c_μ, η_0, and β are model constants equal to 0.71942, 0.71942, 1.42, 1.68, 0.0845, and 0.012, respectively. The other parameters, i.e., $D_{\varepsilon eff}$, G, and S_{ij} are the same as the standard k-ε model. The reader is referred to Yahhot et al. (1992) for further details.

2.7.3 The realizable k-ε model

In the realizable k-ε model, the v_t is calculated as:

$$v_t=c_{\mu r}\frac{k^2}{\varepsilon} \tag{2.21}$$

and k and ε are obtained by solving:

$$\frac{\partial k}{\partial t}+\frac{\partial k u_i}{\partial x_i}-\frac{\partial}{\partial x_i}\left(D_{keff}\frac{\partial k}{\partial x_i}\right)=G-\varepsilon \quad (2.22)$$

$$\frac{\partial \varepsilon}{\partial t}+\frac{\partial \varepsilon u_i}{\partial x_i}-\frac{\partial}{\partial x_i}\left(D_{\varepsilon eff}\frac{\partial \varepsilon}{\partial x_i}\right)=\sqrt{2}c_{1\varepsilon}S_{ij}\varepsilon-c_{2\varepsilon}\frac{\varepsilon^2}{k+\sqrt{\nu\varepsilon}} \quad (2.23)$$

where

$$c_{\mu r}=\frac{1}{A_0+A_sU_s\dfrac{k}{\varepsilon}} \quad (2.24)$$

$$A_s=\sqrt{6}\cos\varphi_s \quad (2.25)$$

$$\varphi_s=\frac{1}{3}arc\cos\left\{\min\left[\max\left(\sqrt{6}W,-1\right),1\right]\right\} \quad (2.26)$$

$$W=\frac{2\sqrt{2}S_{ij}S_{jk}S_{ki}}{S_{mag}S_2} \quad (2.27)$$

$$S_{mag}=\sqrt{S_2} \quad (2.28)$$

where

$$S_2=2\left(dev\left(S_{ij}\right)\right)^2 \quad (2.29)$$

$$U_s=\sqrt{\frac{S_2}{2}+\Omega_{ij}\Omega_{ij}} \quad (2.30)$$

$$\Omega_{ij}=\frac{1}{2}\left(\frac{\partial u_j}{\partial x_i}-\frac{\partial u_i}{\partial x_j}\right) \quad (2.31)$$

$$c_{1\varepsilon}=\max\left(\frac{\eta}{5+\eta},0.43\right) \quad (2.32)$$

$$\eta=S_{mag}\frac{k}{\varepsilon} \quad (2.33)$$

where Ω_{ij} is the vorticity (spin) tensor; σ_k, σ_ε, A_0, and $c_{2\varepsilon}$ are model constants equal to 1, 1.2, 4, and 1.9, respectively. D_{keff}, $D_{\varepsilon eff}$, and G are the same as the RNG k-ε model. The reader is referred to Shih et al. (1995) for further details.

2.7.4 The k-ω model

In the k-ω model, eddy viscosity ν_t is calculated using:

$$\nu_t=\frac{k}{\omega} \quad (2.34)$$

and k and ω are obtained by solving:

$$\frac{\partial k}{\partial t}+\frac{\partial ku_i}{\partial x_i}-\frac{\partial}{\partial x_i}\left(D_{keff}\frac{\partial k}{\partial x_i}\right)=G-c_\mu k\omega \tag{2.35}$$

$$\frac{\partial \omega}{\partial t}+\frac{\partial \omega u_i}{\partial x_i}-\frac{\partial}{\partial x_i}\left(D_{\omega eff}\frac{\partial \omega}{\partial x_i}\right)=\alpha_{k\omega}G\frac{\omega}{k}-\beta\omega^2 \tag{2.36}$$

where

$$D_{keff}=\alpha_k v_t+v \tag{2.37}$$

$$D_{\omega eff}=\alpha_\omega v_t+v \tag{2.38}$$

where ω is the specific dissipation rate, and c_μ, α_k, α_ω, $\alpha_{k\omega}$, and β are the model constants equal to 0.09, 0.5, 0.5, 0.52 and 0.072, respectively (other parameters are calculated the same as the k-ε model). The reader is referred to Wilcox (1988) for further details

2.7.5 The k-ω SST model

The k-ω SST model combines k-ε and k-ω models. In the k-ω SST model:

$$v_t=\frac{a_1 k}{\max\left(a_1\omega, b_1 F_{23}\sqrt{S_2}\right)} \tag{2.39}$$

and k and ω are obtained by solving:

$$\frac{\partial k}{\partial t}+\frac{\partial ku_i}{\partial x_i}-\frac{\partial}{\partial x_i}\left(D_{keffi}\frac{\partial k}{\partial x_i}\right)=\min\left(G, c_1\beta^* k\omega\right)-\beta^* k\omega \tag{2.40}$$

$$\frac{\partial \omega}{\partial t}+\frac{\partial \omega u_i}{\partial x_i}-\frac{\partial}{\partial x_i}\left(D_{\omega eff}\frac{\partial \omega}{\partial x_i}\right)$$

$$=\gamma_i\min\left[S_2, \frac{c_1}{a_1}\beta^*\omega\max\left(a_1\omega, b_1 F_{23}\sqrt{S_2}\right)\right]-\beta\omega^2-(F_1-1)CD_{k\omega} \tag{2.41}$$

where

$$D_{\omega effi}=\alpha_{\omega i}v_t+v \tag{2.42}$$

$$S_2=2S_{ij}S_{ij} \tag{2.43}$$

$$G=2v_t S_{ij}S_{ij} \tag{2.44}$$

$$CD_{k\omega}=2\alpha_{\omega 2}\frac{\partial k}{\partial x_i}\frac{\partial \omega}{\partial x_i}\frac{1}{\omega} \tag{2.45}$$

where the subscript i can be either 1 or 2, depending on the blending functions in the model. α_{k1}, α_{k2}, $\alpha_{\omega 1}$, $\alpha_{\omega 2}$, α_1, b_1, c_1, β^*, γ_1, and γ_2 are model constants equal to 0.85, 1, 0.5, 0.856, 0.31, 1, 10, 0.09, 5/9, and 0.44, respectively, and F_1 and F_{23} are blending functions (nondefined parameters similar to the k-ω model). The reader is referred to Menter (1993) for further details.

2.7.6 The v²-f model

The v^2-f turbulence model is another modified version of the k-ε model, comprising four equations to simulate the effect of turbulence to find a solution for fluid flow motion. This model demonstrates equations for velocity and relaxation factors, along with the general kinetic energy and dissipation equations. It is classified between the RSM and the original k-ε model. A new term v^2 representing velocity, is employed instead of the term for kinetic energy to assess the eddy viscosity (Heyrani et al. 2021). The governing equations of this model are as follows:

$$\frac{\partial k}{\partial t}+\frac{\partial k u_i}{\partial x_i}=P-\varepsilon+\frac{\partial}{\partial x_j}\left(Dk_{eff}\frac{\partial k}{\partial x_j}\right)+S_k \tag{2.46}$$

$$\frac{\partial \varepsilon}{\partial t}+\frac{\partial \varepsilon u_i}{\partial x_i}=\frac{C_{\varepsilon 1}^{'}P-C_{\varepsilon 2}\varepsilon}{T}\frac{\partial}{\partial x_j}\left(Dk\varepsilon_{eff}\frac{\partial \varepsilon}{\partial x_j}\right)+S_\varepsilon \tag{2.47}$$

where

$$\frac{\partial \varepsilon}{\partial t}+\frac{\partial \varepsilon u_i}{\partial x_i}=\frac{C_{\varepsilon 1}^{'}P-C_{\varepsilon 2}\varepsilon}{T}\frac{\partial}{\partial x_j}\left(Dk\varepsilon_{eff}\frac{\partial \varepsilon}{\partial x_j}\right)+S_\varepsilon \tag{2.48}$$

$$\frac{\partial \overline{V^2}}{\partial t}+\frac{\partial \overline{V^2}u_i}{\partial x_i}=kf-6\overline{V^2}\frac{\varepsilon}{k}\frac{\partial}{\partial x_j}\left(Dk_{eff}\frac{\partial \overline{V^2}}{\partial x_i}\right)+S_{\overline{V^2}} \tag{2.49}$$

$$f-L^2\frac{\partial^2 f}{\partial x_i^2}=(C_1-1)\frac{\frac{2}{3}-\frac{\overline{V^2}}{k}}{T}C_2\frac{P}{k}+\left(\frac{5\overline{V^2}/k}{T}\right)+S_f \tag{2.50}$$

and, v_t is calculated using:

$$v_t=c_\mu\overline{v'^2}T \tag{2.51}$$

where, L and T are the length and timescale for turbulence, respectively, and f is the solution to the last equation. The model employs an elliptic operator to compute a term similar to the strain-pressure correlation in RSM. Additionally, C_1, C_2, $C_{\varepsilon 1}^{'}$, and $C_{\varepsilon 2}$ are considered to be constants, and S_k, S_ε, $S_{\overline{V^2}}$, and S_f are source terms. For detailed information on the values of the coefficients, refer to Durbin (1995).

2.8 Large eddy simulation (LES)

Another approach is the Large Eddy Simulation (LES) method, which resolves the filtered NS equations. The method was first introduced by Smagorinsky (1963) and later by Deardorff (1970) for simulation of atmospheric flows in large-scale motions. LES directly resolves the large eddies (those greater than the grid-scale) and models the impact of low-pass filtered residual eddies. The application of the 3D LES method is helpful when the computational grid is fine enough in both the vertical and horizontal scales.

The LES methods are particularly useful when the time-dependent variations of the turbulence structure are important (Rodi et al. 2013). In some LES models, the residual viscosity is a function of kinetic energy of turbulence, and size of the grid, e.g.:

$$v_{SGS}=0.094\cdot\Delta\cdot\sqrt{k} \tag{2.52}$$

where k (m^2/s^2) is the kinetic energy of turbulence, and Δ (m) is the grid scale. It is evident that the amount of SGS viscosity in the flow domain is directly dependent on the size of the model grid; i.e., the finer the grid, the smaller the SGS viscosity. This is a positive attribute of LES models.

A popular subgrid scale model was proposed by Smagorinsky (1963) in which:

$$\nu_{SGS} = C \cdot \Delta^2 \cdot \sqrt{2S_{ij}S_{ij}} \tag{2.53}$$

where C is a constant and the strain rate tensor S_{ij} and Δ are already defined. The reader is referred to Rodi et al. (2013) for further details.

2.9 Detached eddy simulation (DES)

LES methods lead to a high level of instability in complex geometries, and wall function could be violated near the sharp geometric and bathymetric gradients (Spalart 2008). In such cases, the LES methods lead to a high level of computational effort. In some applications, like free surface shallow flows, the situation becomes more problematic, as the horizontal scales are much larger than vertical (depth) scales. Therefore, the fully 3D LES models are not yet practical for large-scale applications.

A compromise is to use RANS to treat near-wall and boundary layer processes and LES elsewhere (i.e., hybrid LES-RANS). This approach is known as Detached Eddy Simulation (DES). DES methods try to benefit from the best aspects of RANS and LES. By combining LES and RANS, the DES methods provide an optimal solution for the complex flow fields. In the DES method, near the wall, a RANS model is applied, and in the domains far enough from the boundaries, the LES model is employed. The resulting solution maintains the accuracy in the majority of the computational domain, and time-dependent features of the flow can also be represented.

A popular DES approach is the $k-\omega$ *SST DES* in which:

$$\frac{D}{Dt}(\rho\omega) = \nabla.(\rho D_\omega \nabla_\omega) + \rho\gamma\frac{G}{\vartheta} - \frac{2}{3}\rho\omega\gamma(\nabla.u) - \rho\beta\omega^2 - \rho(F_1 - 1)CD_{k\omega} + S_\omega \tag{2.54}$$

$$\frac{D}{Dt}(\varphi k) = \nabla.(\rho D_k \nabla_k) + \min\left(\rho G, \left(C_1\beta^*\right)\rho k\omega\right) - \frac{2}{3}\rho k(\nabla.u) - \rho\frac{k^{1.5}}{\bar{d}} + S_k \tag{2.55}$$

where $\bar{d}$ is a length scale defined by:

$$\min\left(C_{DES}\Delta, \frac{\sqrt{k}}{\beta^*\omega}\right) \tag{2.56}$$

with turbulent viscosity defined as:

$$\nu_t = \alpha_1 \frac{k}{\max(\alpha_1\omega, b_1 F_{23} S)} \tag{2.57}$$

Further details can be found in Spalart (1997).

2.10 Impact of buoyancy

The standard k-ε model does not consider the effect of buoyancy on the production of turbulent kinetic energy and its dissipation, and needs to be modified to address this effect. The modification is meant to increase turbulence in the regions of unstable stratification and decrease it in the regions of stable stratification. A common approach is to add source terms to the equations such as k and ε equations. For example, the source term represents turbulence production by buoyancy. A common approach is the simple gradient diffusion hypothesis (SGDH). Nonetheless, this approach leads to underestimation of the term corresponding to buoyancy. The general gradient diffusion hypothesis (GGDH) is an approach that is commonly used to obtain higher precision (Kumar and Dewan 2013; Van Maele and Merci 2006; Worthy et al. 2001).

Yan and Mohammadian (2017) employed the GGDH model. In this model, new terms related to buoyancy are included in the k and ε equations as additional source terms (Daly and Harlow 1970; Rodi 1993; Van Maele and Merci 2006). The resulting buoyancy-modified k-ε model can be written as:

$$\frac{\partial k}{\partial t}+\mu_i\left(\frac{\partial k}{\partial x_i}\right)=\frac{\partial}{\partial x_i}\left(\frac{\nu_t}{\sigma_k}\frac{\partial k}{\partial x_i}\right)+P+G-\varepsilon \tag{2.58}$$

$$\frac{\partial \varepsilon}{\partial t}+\mu_i\left(\frac{\partial \varepsilon}{\partial x_i}\right)=\frac{\partial}{\partial x_i}\left(\frac{\nu_t}{\sigma_\varepsilon}\frac{\partial \varepsilon}{\partial x_i}\right)+C_{1\varepsilon}\frac{\varepsilon}{k}(1-C_{3\varepsilon})(P+G)+C_{2\varepsilon}\frac{\varepsilon^2}{k} \tag{2.59}$$

where the shear production term P and the buoyancy production term G in the k-ε turbulence model in Equations (2.58) and (2.59) are calculated as follows:

$$P=\nu_t\left(\frac{\partial u_t}{\partial x_i}+\frac{\partial u_j}{\partial x_i}\right)\frac{\partial u_t}{\partial x_j} \tag{2.60}$$

$$G=\frac{\overline{\rho' u_i'}}{\rho^2}\left(\frac{\partial P}{\partial x_i}+\rho g_i\right) \tag{2.61}$$

The difference between SGDH and GGDH methods is based on the way the term $\overline{\rho' u_i'}$ is defined. The corresponding equation for SGDH is given by:

$$\overline{\rho' u_i'}=\frac{-\nu_t}{\mathrm{Pr}_t}\frac{\partial \bar{\rho}}{\partial x_j} \tag{2.62}$$

and the GGDH equation can be written as:

$$\overline{\rho' u_i'}=\frac{-3}{2}\frac{C_\mu}{\mathrm{Pr}_t}\frac{k}{\varepsilon}\left(\overline{u_i' u_j'}\frac{\partial P}{\partial x_i}\right)=\frac{-3}{2}\frac{\nu_t}{\mathrm{Pr}_t\,\rho^2 k}\left(\overline{u_i' u_j'}\frac{\partial \bar{\rho}}{\partial x_i}\right) \tag{2.63}$$

The SGDH and GGDH k-ε turbulence models were employed in OpenFOAM to investigate the impact of the buoyancy term, using the following turbulence coefficients $C_\mu=0.09$, $C_{1\varepsilon}=1.44$, $C_{2\varepsilon}=1.92$, as well as a calibrated coefficient $C_{3\varepsilon}=0.6$.

2.11 Summary

The majority of natural and man-made flows are turbulent, and the treatment of turbulence is an essential step in numerically solving the governing equations of fluid flows. The complexity of turbulent motion means that the accurate simulation (DNS) of turbulence is not yet possible in most cases. Because the spectrum of the kinetic energy of turbulence contains wide ranges of turbulent eddies, resolving all these scales requires a significant level of computational effort. Alternatives are the employment and application of turbulence models to include the impacts of turbulent motion in the equations.

The RANS, LES, and DES models are the alternative solutions for turbulence modeling. The RANS-based turbulence models are presently the most popular because they are fast and less complex to implement, and have fewer problems related to instability. Several RANS models have been developed over the past decades, and they have shown reasonable results for engineering applications. However, RANS models do not resolve any scale of turbulent eddies, and in the case where details and behavior of turbulent motion are important, these models would be inapplicable. The LES models resolve major scales of large eddies and model the low pass filtered eddies that are smaller than the size of the numerical grids. As most of the turbulent energy exists in large eddies, the LES models can provide very reasonable results when simulating the energy spectrum of a turbulent motion. However, for the problems with complex geometries, these models are not still applicable and require a high computational cost. The DES-based models are the most recent and hybrid RANS-LES models that benefit from the best aspects of both approaches, and are able to optimize the computing costs and order of accuracy.

This chapter presented a brief explanation of turbulence modeling approaches as well as the most popular models of each approach. The focus of the current chapter was on the RANS models, as they are more frequently utilized in the literature.

References

Daly, B.J. and Harlow, F.H. Transport equations in turbulence. *Phys. Fluid.* 1970, 13, 2634–2649.

Davidson, P.A. *Turbulence: An introduction for scientists and engineers.* Oxford University Press, 2015.

Deardorff, J.W. A numerical study of three-dimensional turbulent channel flow at large Reynolds numbers. *J. Fluid Mech.* 1970, 41, 453–480.

Deltares *Delft3D-FLOW Simulation of Multi-Dimensional Hydrodynamic Flows and Transport Phenomena Including Sediments, User Manual.* Deltares Delft, 2014.

DHI *MIKE 3 Flow Model FM; Hydrodynamic Module Scientific Documentation; Module-User Guide.* Danish Hydraulic Institute, 2017.

Durbin, P.A. Separated flow computations with the k-epsilon-v-squared model. *AIAA J.* 1995, 33, 659–664. doi:10.2514/3.12628.

Elder, J. The dispersion of marked fluid in turbulent shear flow. *J. Fluid Mech.* 1959, 5, 544–560

Harlow F.H. and Welch J.E. Numerical calculation of time-dependent viscous incompressible flow of fluid with free surface, *Physics Fluids* 1965, 8, 2182.

Hervouet, J.-M. TELEMAC, a hydroinformatic system. *La houille blanche* 1999, 3–4, 21–28.

Heyrani, M., Mohammadian, A., Nistor, I., and Dursun, O.F. Numerical modeling of venturi flume. *Hydrology* 2021, 8, 27.

Hirt, C.W., & Nichols, B.D. Volume of fluid (VOF) method for the dynamics of free boundaries. *J. Comput. Physics* 1981, 39(1), 201–225.

Kolmogorov, A.N. Equations of turbulent motion in an incompressible fluid. *Proc. SSSR Acad. Sci.* 1941, 30, 299–303.

Kumar, R., and Dewan, A. Assessment of buoyancy-corrected turbulence models for thermal plumes. *Eng. Appl. Comput. Fluid Mech.* 2013, 7, 239–249.

Launder, B.E., and Spalding, D.B. The numerical computation of turbulent flows. *Comput. Methods Appl. Mech. Eng.* 1974, 3, 269–289.

Menter, F., Zonal two equation k-ω turbulence models for aerodynamic flows. *In* 23rd fluid dynamics, plasmadynamics, and lasers conference. 1993.

Nelson, K. and Fringer, O. Sediment dynamics in wind wave-dominated shallow-water environments. *J. Geophys. Res. Oceans* 2018, 123, 6996–7015.

Prandtl, L. 7. Bericht über Untersuchungen zur ausgebildeten Turbulenz. *ZAMM-J. Appl. Math. Mech.* 1925, 5, 136–139.

Rodi, W. *Turbulence models and their application in hydraulics*. CRC Press, 1993.

Rodi, W. Turbulence modeling and simulation in hydraulics: a historical review. *J. Hydraul. Eng.* 2017, 143, 03117001.

Rodi, W., Constantinescu, G., and Stoesser, T. *Large-Eddy simulation in hydraulics*, 1st edn. CRC Press, 2013.

Shih, T.H., Liou, W.W., Shabbir, A., Yang, Z. and Zhu, J., *A new k-ε eddy viscosity model for high reynolds number turbulent flows. Computers & fluids*, 1995, 24(3), 227–238.

Smagorinsky, J. General circulation experiments with the primitive equations: I. The basic experiment. *Mon. Weather Rev.* 1963, 91, 99–164.

Spalart, P.R. Comments on the feasibility of LES for wings, and on a hybrid RANS/LES approach. *Proceedings of First AFOSR International Conference on DNS/LES*. Greyden Press, 1997.

Spalart, P.R. Detached-Eddy simulation. *Ann. Rev. Fluid Mech.* 2008, 41, 181–202. doi: 10.1146/annurev.fluid.010908.165130.

Van Maele, K. and Merci, B. Application of two buoyancy-modified *k*-ε turbulence models to different types of buoyant plumes. *Fire Safety J.* 2006, 41, 122–138.

Warner, J.C., Sherwood, C.R., Signell, R.P., Harris, C.K., and Arango, H.G. Development of a three-dimensional, regional, coupled wave, current, and sediment-transport model. *Comput. Geosci.* 2008, 34, 1284–1306.

Wilcox, D.C. Reassessment of the scale-determining equation for advanced turbulence models. *AIAA J.* 1988, 46(11), 1299–1310.

Worthy, J., Sanderson, V., and Rubini, P. Comparison of modified *k*-ε turbulence models for buoyant plumes. *Numer. Heat Tr. B: Fund.* 2001, 39, 151–165.

Yan, X. and Mohammadian, A. Numerical modeling of vertical buoyant jets subjected to lateral confinement. *J. Hydraul. Eng.* 2017, 143, 04017016.

Yakhot, VS., Orszag, SA., Thangam, S., Gatski, TB., Speziale, C. Development of turbulence models for shear flows by a double expansion technique. *Physics of Fluids A: Fluid Dynamics*. 1992, 4(7), 1510–20.

Chapter 3

An introduction to OpenFOAM

3.1 OpenFOAM solvers for effluent discharge modeling

The Open source Field Operation And Manipulation (OpenFOAM) code is an object-oriented numerical simulation toolbox for continuum mechanics, written in C++ language, released by OpenCFD Ltd (http://www.openfoam.com). It is capable of supporting all typical features of C++ programming: it enables the construction of new types of data specific to the problem to be solved (i.e., a virtual class for turbulence model with virtual functions such as ε, k, μt, etc.), the bundling of data and operations into hierarchical classes preventing accidental corruption (i.e., a base class for storing mesh data and a derived class for accessing them), a natural syntax for user defined classes (i.e., operator overloading), and it easily permits the re-use of code for equivalent operations on different types of data (i.e., templating) (Jasak, 1996; Jasak et al., 2004; Juretic, 2004).

Let us examine in more in detail the characteristics of OpenFOAM that help computational fluid dynamics (CFD) programmers. First and most importantly is that the toolbox implements operator-based implicit and explicit second and fourth-order finite volume (FV) discretization in three-dimensional space and on curved surfaces. Differential operators can be treated like finite volume calculus (FVC) or finite volume method (FVM) operators.

The first approach performs explicit derivatives that return a field while the second method is an implicit derivation that converts the expression into matrix coefficients. The underlying idea is to think about partial differential equations (PDEs) in terms of a sum of single differential operators that can be discretized separately with different discretization schemes. Differential operators within OpenFOAM are defined as follows:

$$fvm :: ddt = \frac{\partial}{\partial t}$$

$$fvm :: d2dt2 = \frac{\partial^2}{\partial t^2}$$

$$fvm :: div = \sum_i \frac{\partial}{\partial x_i}$$

DOI: 10.1201/9781003181811-3

$$fvm::laplacian = \sum_i \frac{\partial^2}{\partial x_i^2}$$

Building different types of PDEs is now only a matter of combining the same set of basic differential operators in a different way. To give an example of the capability of such a top-level code, consider a standard equation like momentum conservation:

$$\frac{\partial \rho \vec{U}}{\partial t} + \nabla \cdot (\rho \vec{U}\vec{U}) - \nabla \cdot (\mu \nabla \vec{U}) = -\nabla p \quad (3.1)$$

This can be implemented in an astonishingly almost natural language that is ready to be compiled in the C++ source code:

```
  solve
(
      fvm::ddt(rho, U)
    + fvm::div(phi, U)
    - fvm::laplacian(mu, U)
      ==
    - fvc::grad(p)
);
```

This capability lets researchers and engineers concentrate their efforts more on the physics than on programming.

The above example clearly shows that OpenFOAM programmers do not think in terms of cells or faces but in terms of objects (U, rho, phi, etc.) defined as a field of values, no matter the dimension, rank or size, over mesh elements such as points, edges, faces, etc. For example, the velocity field is defined at every cell centroid and boundary-face center, with its given dimensions and the calculated values for each direction, and represented by just a single object U of the class GeometricField.

An important feature allowed by object programming is the dimensional check. The physical quantities of objects are in fact constructed with a reference to their dimensions, and thus, only valid dimensional operations can be performed. Avoiding errors and permitting an easier understanding come directly as a consequence of an easier debug.

OpenFOAM is not really thought of as a ready to use code, but aims to be as flexible as possible in defining new models and solvers in the simplest way. Its strength, in fact, lies in being open, not only in terms of source code but also in its inner structure and hierarchical design, giving the user the opportunity to fully extend its capabilities. Moreover, the possibility of using top-level libraries containing a set of models for the same purpose, which refer to the same interface, guarantees programmers smooth and efficient integration with the built-in functionality.

Most of the selections necessary to set up calculations are done at runtime, meaning that options can change while the code is running.

OpenFOAM consists of a library of efficient CFD-related C++ modules. These features can be combined together to create "Solvers" and "Utilities" which are listed below.

- **OpenFOAM Solvers:**
  - Basic CFD
  - Incompressible Flows
  - Compressible Flows
  - Multiphase Flows
  - DNS and LES
  - Combustion
  - Heat Transfer
  - Electromagnetics
  - Solid Dynamics
  - Finance
- **OpenFOAM Utilities:**
  - Preprocessing
  - The FoamX Case Manager
  - Other Preprocessing Utilities
  - Post-processing
  - The paraFoam Post-processor
  - Third-party Post-processing
  - Other Post-processing Utilities
  - Mesh Processing
  - Mesh Generation
  - Mesh Converts
  - Mesh Manipulation
- **OpenFOAM Libraries:**
  - Model Libraries
  - Turbulence
  - Large-eddy Simulation (LES)
  - Transport Models
  - Thermophysical Models
  - Chemical Kinetics
- **Other Features:**
  - Linear System Solvers
  - ODE System Solvers
  - Parallel Computing
  - Mesh Motion
  - Numerical Method

3.1.1 Model preparation

In the OpenFOAM tool-chain, it is possible to use an entirely open source software (OSS) tool-chain, or a combination of OSS and commercial tools. Figures 3.1 and 3.2 show the simulation process priority and the whole tool-chain of OpenFOAM, respectively.

The aim of this section is to explain how OpenFOAM's existing applications are used in order to simulate the desired positively/negatively buoyant jets.

The preparation of the model is divided into two parts: the first step is to implement the required equations in the basic solver, and the second step involves setting up the case and the input files.

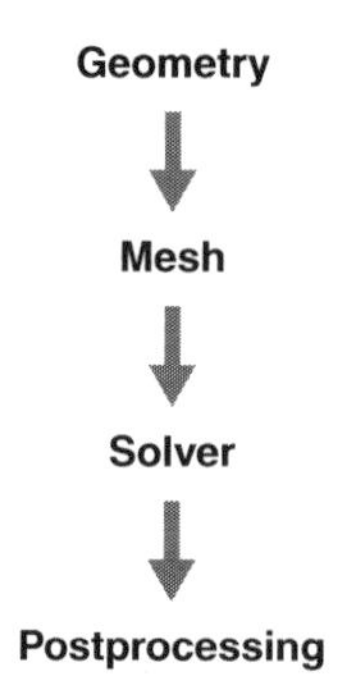

Figure 3.1 Simulation process priority in OpenFOAM.

The present chapter introduces two OpenFOAM solvers for effluent discharge modeling. The first one is based on the standard pisoFoam solver within OpenFOAM and is extended to solve the transport equations. The temperature and concentration evolutions are modeled using the advection-diffusion equation as:

$$\frac{\partial T}{\partial t}+u\frac{\partial T}{\partial x}+v\frac{\partial T}{\partial y}+w\frac{\partial T}{\partial z}=k_{eff}\left(\frac{\partial^2 T}{\partial x^2}+\frac{\partial^2 T}{\partial y^2}+\frac{\partial^2 T}{\partial z^2}\right) \tag{3.2}$$

$$\frac{\partial C}{\partial t}+u\frac{\partial C}{\partial x}+v\frac{\partial C}{\partial y}+w\frac{\partial C}{\partial z}=D\left(\frac{\partial^2 C}{\partial x^2}+\frac{\partial^2 C}{\partial y^2}+\frac{\partial^2 C}{\partial z^2}\right) \tag{3.3}$$

with

$$k_{eff}=\frac{v_t}{Pt_r}+\frac{v}{P_r} \tag{3.4}$$

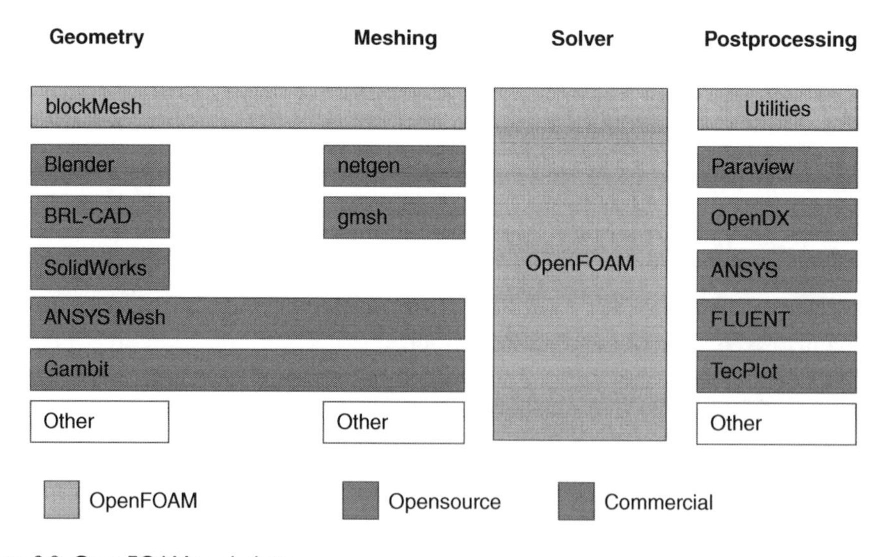

Figure 3.2 OpenFOAM tool-chain.

where T is the fluid temperature, k_{eff} is the heat transfer coefficient, Pt_r is the turbulent Prandtl number, and P_r is the Prandtl number.

The densimetric Froude number Fr is one of the most important parameters in characterizing the flow. The Fr number is the ratio of inertia to buoyancy force, which can be expressed as:

$$Fr = \frac{W_j}{\sqrt{g'D}} \tag{3.5}$$

with

$$g' = g\frac{\rho_a - \rho_j}{\rho_a} \tag{3.6}$$

where W_j is the initial velocity, D is the diameter of the discharge port, g is the gravitational acceleration, ρ_a is the ambient density, and ρ_j is the jet's initial density.

The density difference, which is incorporated in the densimetric Froude number, produces the buoyancy. Liquid wastes discharged from industrial outfalls are categorized into two major classes based on their density. One type is the effluent that has a higher density than the ambient water body. In this case, the discharged effluent has a tendency to sink as a negatively buoyant jet. The second type is the effluent that has a lower density than the ambient water body and is hence defined as a (positively) buoyant jet, which causes the effluent to rise. The density can be calculated based on the equation of state of seawater of Millero and Poisson (1981).

The governing equations are the well-known Navier-Stokes equations for three-dimensional, incompressible fluids as follows:

Continuity equation:

$$\frac{\partial u}{\partial x} + \frac{\partial v}{\partial y} + \frac{\partial w}{\partial z} = 0 \tag{3.7}$$

Momentum equations:

$$\begin{aligned}\frac{\partial u}{\partial t} + u\frac{\partial u}{\partial x} + v\frac{\partial u}{\partial y} + w\frac{\partial u}{\partial z} &= -\frac{1}{\rho}\frac{\partial P}{\partial x} + \frac{\partial}{\partial x}\left(\upsilon_{eff}\left(\frac{\partial u}{\partial x}\right)\right) \\ &+ \frac{\partial}{\partial y}\left(\upsilon_{eff}\left(\frac{\partial u}{\partial y}\right)\right) + \frac{\partial}{\partial z}\left(\upsilon_{eff}\left(\frac{\partial u}{\partial z}\right)\right)\end{aligned} \tag{3.8}$$

$$\begin{aligned}\frac{\partial v}{\partial t} + u\frac{\partial v}{\partial x} + v\frac{\partial v}{\partial y} + w\frac{\partial v}{\partial z} &= -\frac{1}{\rho}\frac{\partial P}{\partial y} + \frac{\partial}{\partial x}\left(\upsilon_{eff}\left(\frac{\partial v}{\partial x}\right)\right) + \frac{\partial}{\partial y}\left(\upsilon_{eff}\left(\frac{\partial v}{\partial y}\right)\right) \\ &+ \frac{\partial}{\partial z}\left(\upsilon_{eff}\left(\frac{\partial v}{\partial z}\right)\right) - g\frac{\rho - \rho_0}{\rho}\end{aligned} \tag{3.9}$$

$$\begin{aligned}\frac{\partial w}{\partial t} + u\frac{\partial w}{\partial x} + v\frac{\partial w}{\partial y} + w\frac{\partial w}{\partial z} &= -\frac{1}{\rho}\frac{\partial P}{\partial z} + \frac{\partial}{\partial x}\left(\upsilon_{eff}\left(\frac{\partial w}{\partial x}\right)\right) \\ &+ \frac{\partial}{\partial y}\left(\upsilon_{eff}\left(\frac{\partial w}{\partial y}\right)\right) + \frac{\partial}{\partial z}\left(\upsilon_{eff}\left(\frac{\partial w}{\partial z}\right)\right)\end{aligned} \tag{3.10}$$

where u, v, and w are the mean velocity components in the x, y, and z direction, respectively, t is the time, P is the fluid pressure, υ_{eff} represents the effective kinematic viscosity ($\upsilon_{eff} = \upsilon t + \upsilon$), υt is the turbulent kinematic viscosity, g is the gravitational acceleration, ρ is the fluid density, and ρ_0 is the reference fluid density.

One should note that the equations are divided by density (ρ), and the buoyancy term is added to the momentum equation in the vertical direction (y-coordinate) to account for variable density effects.

As can be observed from the governing equations, the velocity cannot be calculated until the pressure is solved, and the pressure cannot be computed until the velocity is known. The pisoFoam solver starts by estimating the pressure, and then uses this estimated pressure to calculate an intermediate velocity field and the mass fluxes at the cells' faces. With this information, it then solves the pressure equation. The fluxes are then corrected to satisfy continuity, velocities are corrected on the basis of the new pressure field, and the transport equation is solved using the finite volume method.

The temporal term may be discretized by first-order implicit Euler scheme. The advection and diffusion terms are discretized by the standard finite-volume method using Gaussian integration with a linear interpolation scheme for calculating values at face centers from cell centers. For the pressure field, the preconditioned conjugate gradient (PCG) method is used to solve the linear system. The preconditioned bioconjugate gradient (PBiCG) method may be used for other fields: U, T, C, k, ε, and ω. In order to enhance the rate of convergence for iterative solvers, the diagonal incomplete Cholesky (DIC) preconditioner is used to calculate the pressure field. This is a simplified diagonal-based preconditioner for symmetric matrices. The diagonal incomplete lower upper (DILU) preconditioner is used for the other fields: U, T, C, k, ε, and ω, which mostly include asymmetric matrices to be solved.

Another solver that is popular for discharge modeling is the standard twoLiquidMixingFoam solver in OpenFOAM, which is an incompressible multiphase solver for miscible fluids. The governing equations for this solver can be expressed as:

$$\nabla \cdot U = 0 \tag{3.11}$$

$$\frac{\partial \rho U}{\partial t} + \nabla \cdot (\rho U U) = -\nabla \cdot (p_{rgh}) - gh\nabla\rho + \nabla \cdot (\rho T) \tag{3.12}$$

with

$$\rho = \alpha_1\rho_1 + \alpha_2\rho_2 = \alpha_1\rho_1 + (1-\alpha_1)\rho_2 \tag{3.13}$$

$$T = -\frac{2}{3}\bar{\mu}_{eff}\nabla \cdot \mathrm{UI} + \bar{\mu}_{eff}\nabla U + \bar{\mu}_{eff}(\nabla U)^{\mathrm{T}} \tag{3.14}$$

$$\bar{\mu}_{eff} = \alpha_1(\mu_{eff})_1 + \alpha_2(\mu_{eff})_2 \tag{3.15}$$

$$(\mu_{eff})_{\mathrm{i}} = (\mu - \mu_{\mathrm{t}})_{\mathrm{i}} \tag{3.16}$$

where t is time, U is velocity, ρ is density, p_{rgh} is dynamic pressure, p represents pressure, and $\nabla \cdot (p_{rgh})$ and $gh\nabla\rho$ are obtained by using $P = p_{rgh} + \rho gh$, g is gravitational acceleration, α is volume fraction, μ is dynamic viscosity, and μt is turbulent viscosity, and subscript i denotes either fluid 1 or 2. In such simulations, fluids 1 and 2 represent the jet and ambient water, respectively.

The alpha diffusion equation is given by:

$$\frac{\partial \alpha_1}{\partial t} + \nabla \cdot (U\alpha_1) = \nabla \cdot \left(\left(D_{ab} + \frac{v_t}{S_C} \right) \nabla \alpha_1 \right) \tag{3.17}$$

where D_{ab} is the molecular diffusivity, v_t is the turbulent eddy viscosity, and S_C is the turbulent Schmidt number.

The continuity, momentum, and alpha diffusion equations can be solved using the finite volume method by using twoLiquidMixingFoam within the OpenFOAM framework. This solver has been widely employed and validated in previous studies (e.g., Gildeh et al., 2021).

3.2 Mesh generation in OpenFOAM

3.2.1 *Basic steps of mesh generation in OpenFOAM*

Mesh generation may not seem an easy task in OpenFOAM at first glance. It is in fact a common challenge in all CFD models. Creating a suitable mesh that characterizes the geometry of the problem and helps with numerical stability is the first and most important step in building a numerical model. Creating a mesh in OpenFOAM starts with the following steps:

1 Label the coordinate system: draw a three-dimensional Cartesian coordinate system, with the vertical direction as the y-axis.
2 Draw geometric figures: first draw the upper part and then the lower part.
3 Add basic auxiliary lines: it is necessary to ensure that each line runs through; if not, auxiliary lines need to be added.
4 Bold vertices: all intersections are vertices.
5 Determine the blocks: improve the auxiliary line; the numbering starts from 0, such as B0, B1, B2, etc.; check whether the block element is perfect.
6 Coding vertices: all vertices are numbered and must be coded in a reasonable order. The coding involving the order in OpenFOAM is advanced in the order of x, y, and z directions.

An example of a mesh generated by following the above steps is shown in Figure 3.3.

The specific process is as follows:

1 First, taking the above sketch as an example, enter the vertex information in the order of x, y, and z (assume a small value for z such as 0.1).
2 Enter the blocks information and provide the information for the eight vertices. The number of x, y, and z meshes are set to 15, 15, and 1, respectively, and the ratio of adjusting the mesh size is (1 1 1).

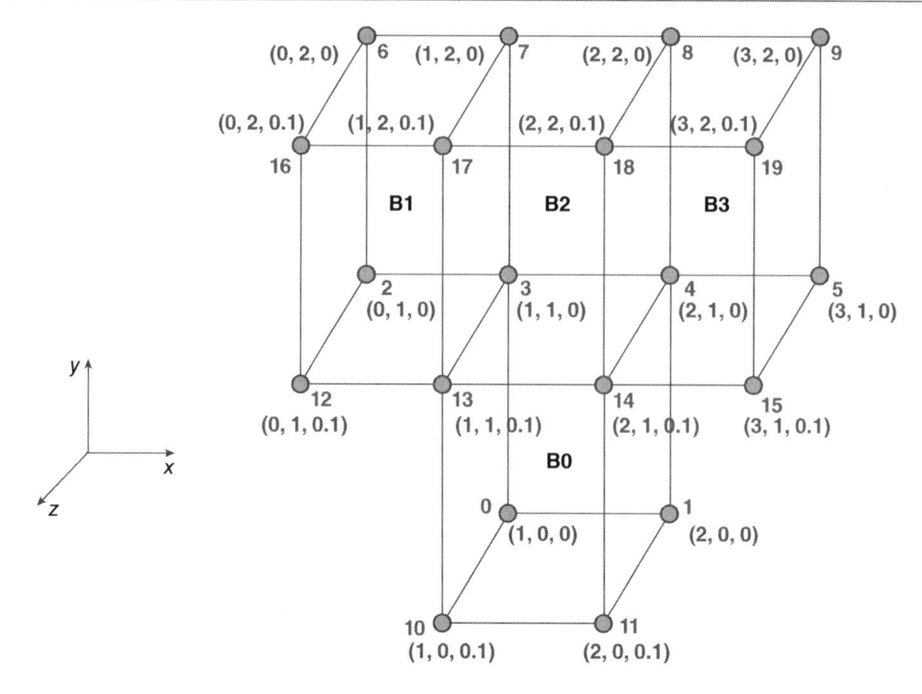

Figure 3.3 Example of input sketch for use in OpenFOAM.

3 Enter the boundary information and name it. The open boundary between the entrance and the exit is represented by patch; the wall is named fixedWalls, which is directly represented by wall; the front and rear surfaces are named frontAnd-Back, which is represented by empty.

```
Vertices                // enter the vertex information in the order
of x, y, and z.
(
        (1 0 0) //0
        (2 0 0) //1
        (0 1 0) //2
        (1 1 0) //3
        (2 1 0) //4
        (3 1 0) //5
        (0 2 0) //6
        (1 2 0) //7
        (2 2 0) //8
        (3 2 0) //9
        (1 0 0.1) //10
        (2 0 0.1) //11
        (0 1 0.1) //12
        (1 1 0.1) //13
        (2 1 0.1) //14
        (3 1 0.1) //15
        (0 2 0.1) //16
        (1 2 0.1) //17
```

```
        (2 2 0.1) //18
        (3 2 0.1) //19
        );
blocks                          // Enter the blocks information
(
         hex (0 1 4 3 10 11 14 13) (15 15 1) simpleGrading
(1 1 1)
         hex (2 3 7 6 12 13 17 16) (15 15 1) simpleGrading
(1 1 1)
         hex (3 4 8 7 13 14 18 17) (15 15 1) simpleGrading
(1 1 1)
         hex (4 5 9 8 14 15 19 18) (15 15 1) simpleGrading
(1 1 1)
);
edges
(
);
Boundary                 // Enter the boundary information and name it

(
        Inlet
        {
             type patch;             // The open boundary between the
entrance and the exit is represented by patch faces
             (
                      (0 1 11 10)
             );
        }
        outlet                  //
        {
                type patch;
                faces
                (
                      (6 7 17 16)
                      (7 8 18 17)
                      (8 9 19 18)
                 );
        }
        fixedWalls                        // the wall is named
fixedWalls, which is directly represented by wall
                                               faces
        {
                type wall;
                (
                      (0 3 13 10)
                      (2 3 13 12)
                      (2 6 16 12)
                      (1 4 14 11)
                      (4 5 15 14)
                      (5 9 19 15)
                );
        }
```

```
        frontAndBack               // the front and rear surfaces are
named frontAndBack, which is represented by empty.
        {
                type empty;
                faces
                (
                        (0 1 4 3)
                        (2 3 7 6)
                        (3 4 8 7)
                        (4 5 9 8)

                        (10 11 14 13)
                        (12 13 17 16)
                        (13 14 18 17)
                        (14 15 19 18)
                );
        }
);

mergePatchPairs
(
);
```

3.2.2 Common mesh generation methods

Mesh generation is one of the initial tasks that a CFD modeler should complete before solving the problem at hand. The mesh system should capture the geometry and topology of the problem as well as provide the computational cells to solve the mathematical equations. There are different methods in mesh generation in OpenFOAM. The following describes four of them.

1. Cartesian grid generation method
   One can approximate curvilinear geometry of complex geometry in Cartesian coordinates by step-by-step approximation. But, this method takes a lot of time and is very tedious to use. Although it is possible to refine the mesh, the stepwise approximation is not smooth and thus leads to large errors, which waste computational resources. Therefore, there are limitations to using CFD methods based on simple coordinate systems (Cartesian or cylindrical), as these systems fail when modeling complex geometries such as airfoil shapes.
2. Structured grid generation method
   In this method, elements are aligned in a specific way, or they follow a structured pattern, thus it is called a structured grid. The intersection of coordinate lines identifies the grid points, which then can be arranged in an array.

   The advantage of the structured grid is that the geometric information of each node and the control volume must be stored, but the neighbor relationship of the node can be automatically obtained according to the grid numbering rule. Therefore, the data structure is simple, and it is not necessary to store this kind of information, which is a major advantage of structured grids. In addition, it also has the following advantages: a) the grid generation speed is fast; b) the grid generation quality is good; c) most of the fitting of the surface or space is obtained by parametric or spline interpolation (i.e., the resulting area is smooth) and it is easier to approach the actual model. It can easily realize the boundary fitting of the region, and is suitable for the calculation of fluid and surface stress concentration.

However, the structured grid also has certain disadvantages: the scope of application is relatively narrow and it is only suitable for graphics with regular shapes.

Structured methods are further divided into single-block and multiblock methods:

1 Structured monolithic grid
A single block is used to cover the entire geometry. It is used for simple shapes, where a single block is sufficient to cover the domain.
2 Structured multiblock grid
In structured meshing, another method used to mesh complex geometry is called the multiblock method. Suppose that in the core of a cylinder, users want to generate a finer mesh or a different pattern than the outer regions; in order to separate these regions, sometimes multiple bounding boxes are used. Similarly, for complex geometries, it is not possible to cover shapes with a single bounding box, so different shapes in the same CFD space generate different bounding boxes, and these blocks can then be combined to form a single structured net grid. Since the mesh generation process involves multiple blocks, it is called a multiblock structured mesh.
3 Unstructured mesh generation method
In this method, all the discrete cells generated in the CFD space are randomly arranged and do not follow any single pattern, as the name "unstructured grid" suggests. Unstructured mesh means that the interior points in the mesh area do not have the same adjacent cells, which can be of various shapes such as tetrahedron, prism, or hexahedron. The number of meshes connected to different interior points within the meshed region varies.
The advantage of the unstructured mesh generation method is that it has no regular topology and no concept of layers. The distribution of grid nodes is arbitrary and therefore flexible. However, larger memory capacity is required for computation.
4 Hybrid mesh generation method
Hybrid meshes are the most complex meshing processes that are used to study complex geometries. Because the geometry is complex and the outline of the shape cannot be captured by using a single element type, multiple types are available such as prisms, wedges, and hexahedrons. Similarly, for volume meshes of spheres along the surface of a sphere, tetrahedra are used, while hexahedra are used internally. Therefore, the combination of different shapes provides a hybrid mesh.

3.2.3 Parameter definition

There are various parameters in OpenFOAM that are utilized in the blockMeshDict file as described below.

1 Vertices parameter definition
The vertices parameter lists all the point coordinates contained in the block. The point number starts from 0.

2 Blocks parameter definition
The blocks parameter settings include shape flags and divisions. The shape flags are basically hex because the block is generally the sequence number of the 8 points required by the block specified in the first () after the hexahedron hex, and arranged as shown in the block coding sequence diagram. The second () is used to define the number of mesh divisions in the xyz direction. blockMesh supports two division methods simpleGrading and edgeGrading.
simpleGrading () specifies the use of uniform scaling in three directions, and specifies the size of these three proportions respectively. For example, the simple-Grading (2 3 4) setting means that the scale factors along the x, y, and z directions are 2, 3, and 4, respectively.
edgeGrading () gives the complete cell scale for each boundary, from the first voxel to the last voxel direction.
For example, simpleGrading (2 2 2 2 3 3 3 3 4 4 4 4); this setting means that the scale along sides 0~3 is 2, the scale along sides 4~7 is 3, and the scale along sides 8~11 is 4, and the effect is the same as using simpleGrading above.
3 Edges parameter definition
By default, the edge connecting two vertices is straight, but when the edge is a curve, it can be specified by an entry in a list named edges. This list can be ignored if the geometry space does not include curves. In Section 3.2.1, the edge connecting any two points is a straight line, so the list is ignored in the definition. Common boundary types in OpenFOAM are shown in Table 3.1.
Consider the following example:

```
edges
(
arc 1 5 (1.1 0.0 0.5)
);
```

This means that the keyword arc is followed by the label numbers 1 and 5 of the two vertices connected by the edge. When using the arc keyword to specify an arc, a point through the circle where the arc is located must be specified; then the arc passes through the internal point (1.1, 0.0, 0.5).
4 Boundary parameter definition
This section is used to define the name and type of the mesh boundary, as well as a list of vertex numbers for each face (numbering order satisfies the right-hand criterion). The name of the user-defined boundary, and the type of boundary

Table 3.1 Common boundary types in OpenFOAM

Keyword	*Description*	*Parameters to be supplemented*
Arc	Connect using arcs	A point through which the arc passes
simpleSpline	Connect using splines	A series of interior points
polyLine	Connect using a series of straight lines	A series of interior points
polySpline	Connect using a series of splines	A series of interior points
Line[a]	Straight connection	-

Note:
[a] The line keyword was not originally required, but it was added to the keyword list for completeness.

can be specified by the parameter type. The boundary types include: wall (wall), symmetry plane (symmetryPlane), periodic boundary (cyclic), inconsistent periodic boundary (cyclicAMI), two-dimensional boundary axisymmetric boundary (wedge), and 2D boundary (empty).

When using some specific boundary types, attention must be paid to their usage scenarios and associated files. Currently, OpenFOAM only supports 3D grid calculation. If users want to calculate 2D problems, they can set both side walls in the 3D calculation domain as empty boundaries, which are used as 2D calculation boundaries. The faces parameter specifies the order of the points that make up the boundary face, which conforms to the right-hand rule.

5 mergePatchPairs parameter definition

blockMesh allows users to create meshes using multiple blocks. In the case of including multiple blocks, it is necessary to deal with the connection problem between each block. There are two methods for merging as described in the following.

i Face matching

It is required that if block A and block B are to be merged, the vertices of the interface patch A1 and patch B1 are exactly the same. When connecting blocks by face matching, the user does not need to define parameters within mergePatchPairs. blockMesh will automatically match these two patches into inner faces, as shown in the following example:

```
mergePatchPairs
(
);
```

ii Face fusion

There is a mapping relationship between the two patches in two blocks, and the vertices are not required to be exactly the same. The fusion rules are as follows.

- The masterPatch on the main surface remains unchanged, and the coordinates of all points on it remain unchanged.
- If there is a gap between the masterPatch on the main surface and the slavePatch on the secondary surface, project the slavePatch on the secondary surface onto the masterPatch on the main surface to meet the surface fusion requirements.
- Adjust the position of the nodes on the secondary surface through the minimum tolerance value, improve the node matching degree between the primary surface and the secondary surface, and remove the subtle edges smaller than the minimum tolerance.
- When the main and auxiliary surfaces partially overlap, the overlapping part will become an internal surface, and the nonoverlapping part will still be an external surface, and boundary conditions need to be defined.

If the secondary surface slavePatch is fully integrated into the main surface, the secondary surface will be removed. When connecting blocks by face fusion, the two patches to be merged in the mergePatchPairs parameter must be defined, as shown in the following example:

```
mergePatchPairs
(
(<masterPatch> <slavePatch>)
);
```

As seen above, and in general, the grid generation is one of the most challenging tasks in creating a CFD model and OpenFOAM is not an exception. Section 3.5 reviews different utilities in OpenFOAM for mesh generation to solve the discharge mixing problems.

Now that we have discussed available solvers in OpenFOAM and mesh generation and its importance in solving the effluent discharge problems, and before moving to Section 3.4 to discuss the post processing of the results, we will review an example of an effluent discharge model in pisoFoam solver.

3.3 Effluent discharge model preparation in OpenFOAM using pisoFoam solver

3.3.1 PISO algorithm

The governing equations are solved numerically using the finite volume method. The solver, which is used within OpenFOAM, is the modified pisoFoam. This solver is a transient solver for incompressible flow. The code first predicts the velocity field by solving the velocity equation (momentum equations). Although the continuity equation is presented, pisoFoam does not actually solve it; instead, it solves a pressure Poisson equation in PISO (Pressure-Implicit with Splitting of Operators). Rather than solve all of the coupled equations in a coupled or iterative sequential fashion, PISO splits the operators into an implicit predictor and multiple explicit corrector steps. At each time step, velocity, concentration, and temperature are predicted, and then pressure and velocity are corrected. The velocity is predicted implicitly because of the greater stability of implicit methods, which means that a set of coupled linear equations, expressed in matrix-vector form as Ax = b, are solved. It then solves the concentration and temperature equations (Equations 3.2 and 3.3). As in the velocity predictor, the concentration and temperature are predicted implicitly from Equations 3.2 and 3.3.

In order to solve the equations, a new solver, called mypisoFoam, was developed and added to OpenFOAM. Here, the implementation of the momentum equation is shown in mypisoFoam solver as an example.

As previously explained, each term in a PDE is represented in OpenFOAM code using the classes of static functions finiteVolumeMethod and finiteVolumeCalculus, shortened as fvm and fvc, respectively. fvm and fvc contain static functions, representing deferential operators that discretize the terms in the PDE.

Equations, and terms of equations are declared as tmp<Type> where <Type> is either <fvVectorMatrix> if the equation is a vector equation, like the momentum equation, or <fvScalarMatrix> if the equation is a scalar equation, like the ε-equation. The names indicate that the resulting discretized equations are stored as matrices.

The first part of the code for momentum equations (Equations 3.8–3.10) is to introduce a new definition of density that is a function of both temperature and salinity. The density calculation is based on Millero and Poisson (1981) as written in Equations 3.18–3.21 below.

$$\rho = f(S,T) \tag{3.18}$$

Here, the density is calculated for both the jet and the ambient water according to the equation of state of seawater proposed by Millero and Poisson (1981):

$$\rho = \rho_t + AS + BS^{3/2} + CS \tag{3.19}$$

where

$$\begin{aligned} A &= 8.24493 \times 10^{-1} - 4.0899 \times 10^{-3} T + 7.6438 \times 10^{-5} T^2 - 8.2467 \times 10^{-7} T^3 \\ &\quad + 5.3875 \times 10^{-9} T^4 \\ B &= -5.72466 \times 10^{-3} + 1.0227 \times 10^{-4} T - 1.6546 \times 10^{-6} T^2 \\ C &= 4.8314 \times 10^{-4} \end{aligned} \tag{3.20}$$

and ρ_t is the density of water that varies with the temperature as follows:

$$\begin{aligned} \rho_t &= 999.842594 + 6.793952 \times 10^{-2} T - 9.095290 \times 10^{-3} T^2 \\ &\quad + 1.001685 \times 10^{-4} T^3 - 1.120083 \times 10^{-6} T^4 + 6.536336 \times 10^{-9} T^5 \end{aligned} \tag{3.21}$$

This has been implemented in the main solver as follows:

```
rho==1000;
rho==(rho/1000*(999.842594+6.793952e-2*T-(9.095290e-3*pow(T,2))+
(1.001685e-4*pow(T,3))-(1.120083e-6*pow(T,4))+(6.536336e-9*pow(T,5))+
S*(8.24493e-1-4.0899e-3*T+(7.6438e-5*pow(T,2))-(8.2467e-7*pow(T,3))+
(5.3875e-9*pow(T,4)))+(pow(S,1.5))*(-5.72466e-3+1.0227e-4*T-1.6546e-6*pow(T,2))+S*(4.8314e-4)));
```

The second part of the UEqn.H implements the LHS of the momentum equations (Equations 3.8–3.10).

```
///
    fvVectorMatrix UEqn
    (
        fvm::ddt(U)
      + fvm::div(phi, U)
      + turbulence->divDevReff(U)
     - g*(rho-1000)/1000

    );
///

            UEqn.relax();
```

And finally, the last part of the algorithm adds the RHS (which is equal to the pressure gradient) and solves the momentum equations.

```
if (momentumPredictor)
{
    solve(UEqn == -fvc::grad(p));
}
```

The PISO algorithm, which is used in the mypisoFoam solver can be understood, for simplicity, by considering a one-dimensional, inviscid flow along the x-direction. So, the momentum equation is simplified to:

$$\frac{\partial u}{\partial t} + u\frac{\partial u}{\partial x} = -\frac{1}{\rho}\frac{\partial P}{\partial x} \tag{3.22}$$

By using the Euler implicit time stepping with linear interpolation of values to the cell faces and linearization of the convective term by taking the convective velocity from the old time step n, the discretized implicit velocity predictor forms the following equation:

$$\left[\frac{1}{\Delta t}+\left(\frac{u^n_{i+\frac{1}{2}}-u^n_{i-\frac{1}{2}}}{2\Delta x}\right)\right]\Delta V u_i^* +\left(\frac{u^n_{i+\frac{1}{2}}}{2\Delta x}\right)\Delta V u_{i+1}^* -\left(\frac{u^n_{i-\frac{1}{2}}}{2\Delta x}\right)\Delta V u_{i-1}^* = \frac{u_i^n}{\Delta t}\Delta V-\left(\frac{1}{\rho}\frac{\partial P}{\partial x}\right)_i^n \Delta V \tag{3.23}$$

where the predicted values are denoted by *. Notice that pressure is taken from the old time step n since it is yet unknown. Now, the cell volume ΔV can be divided out as follows to get the correct coefficient matrices and vectors that are used in the corrector step:

$$\left[\frac{1}{\Delta t}+\left(\frac{u^n_{i+\frac{1}{2}}-u^n_{i-\frac{1}{2}}}{2\Delta x}\right)\right]u_i^* +\left(\frac{u^n_{i+\frac{1}{2}}}{2\Delta x}\right)u_{i+1}^* -\left(\frac{u^n_{i-\frac{1}{2}}}{2\Delta x}\right)u_{i-1}^* = \frac{u_i^n}{\Delta t}-\left(\frac{1}{\rho}\frac{\partial P}{\partial x}\right)_i^n \tag{3.24}$$

In vector form, this becomes:

$$Cu^* = r - \nabla P^n \tag{3.25}$$

where C is the coefficient array multiplying the solution u^* vector and r is the right-hand side explicit terms. If the viscous and turbulent stress terms are included, they would modify the coefficient matrix C and would not change the general form of the matrix-vector equation. This equation can be changed to:

$$Au^* + H'u^* = r - \nabla P^n \tag{3.26}$$

where A is the diagonal matrix of C and H' is the off-diagonal matrix of A (i.e., $A + H' = C$). Using a matrix solver, the above equation is solved for the predicted velocity u^*.

Moreover, the discretized explicit velocity corrector is written as:

$$\left[\frac{1}{\Delta t}+\left(\frac{u^n_{i+\frac{1}{2}}-u^n_{i-\frac{1}{2}}}{2\Delta x}\right)\right]u_i^{**} +\left(\frac{u^n_{i+\frac{1}{2}}}{2\Delta x}\right)u_{i+1}^* -\left(\frac{u^n_{i-\frac{1}{2}}}{2\Delta x}\right)u_{i-1}^* = \frac{u_i^n}{\Delta t}-\left(\frac{1}{\rho}\frac{\partial P}{\partial x}\right)_i^* \tag{3.27}$$

The first corrected velocity u^{**} is being solved from the predicted velocity u^*, old velocity u^n, and the first corrected pressure P^*. The problem is that the corrected pressure is still unknown. This equation can be expressed in matrix-vector form, as in Equation (3.26):

$$Au^{**} + H'u^* = r - \nabla P^* \tag{3.28}$$

Introducing $H = r - H'u^*$ and inverting A (which is easy since it is diagonal), Equation (3.28) becomes:

$$u^{**} = A^{-1}H - A^{-1}\nabla P^* \tag{3.29}$$

The point of the corrector step is to make the corrected velocity field divergence free so that it adheres to the continuity equation. By applying the divergence to the above equation and recognizing that $\nabla u^{**} = 0$ due to the continuity equation, this yields a Poisson equation for the first corrected pressure:

$$\nabla^2\left(A^{-1}p^*\right) = \nabla\cdot\left(A^{-1}H\right) \tag{3.30}$$

When the first corrected pressure p* has been calculated, Equation (3.30) can be solved for the first corrected velocity u^{**}.

The higher correction steps might be applied using the same A matrix and H vector. The second correction step is also shown below:

$$\nabla^2\left(A^{-1}p^{**}\right) = \nabla\cdot\left(A^{-1}H\right) \tag{3.31}$$

$$u^{***} = A^{-1}H - A^{-1}\nabla P^{**} \tag{3.32}$$

where p^{**} and u^{***} are the second corrected pressure and velocity, respectively. This method works for the other implicit time stepping schemes, for instance Crank-Nicholson or second-order backward. Issa (1985) states that if a second order accurate time stepping scheme is used, then three corrector steps should be used to reduce the discretization error due to PISO algorithm to second-order.

The mypisoFoam solver only solves continuity and momentum equations; hence, we added the advection-diffusion equation for concentration and temperature to this solver and compiled it to use for our case, as explained as follows.

3.3.2 *A new solver is born*

The PISO loop within the main solver is shown in the following.

```
// --- PISO loop

for (int corr=0; corr<nCorr; corr++)
{
    volScalarField rAU(1.0/UEqn.A());

    U = rAU*UEqn.H();
    phi = (fvc::interpolate(U) & mesh.Sf())
        + fvc::ddtPhiCorr(rAU, U, phi);

    adjustPhi(phi, U, p);

    // Non-orthogonal pressure corrector loop
    for (int nonOrth=0; nonOrth<=nNonOrthCorr; nonOrth++)
    {
        // Pressure corrector

        fvScalarMatrix pEqn
        (
            fvm::laplacian(rAU, p) == fvc::div(phi)
        );

        pEqn.setReference(pRefCell, pRefValue);

        if
        (
            corr == nCorr-1
         && nonOrth == nNonOrthCorr
        )
        {
            pEqn.solve(mesh.solver("pFinal"));
        }
        else
        {
            pEqn.solve();
        }
```

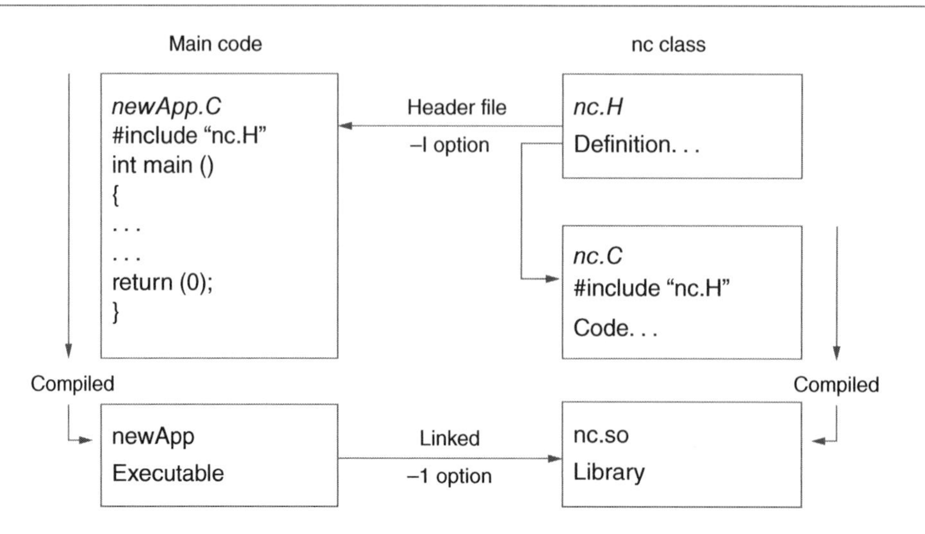

Figure 3.4 Header files, source files, compilation, and linking.

Compilation is an integral part of application development that requires careful management since every piece of code requires its own set of instructions to access dependent components of the OpenFOAM library. In UNIX/Linux systems, these instructions are often organized and delivered to the compiler using the standard UNIXmake utility. OpenFOAM, however, is supplied with the wmake compilation script that is based on make, but is considerably more versatile and easier to use; wmake can be used on any code, not only the OpenFOAM library. To understand the compilation process, we first need to explain certain aspects of C++ and its file structure, shown schematically in Figure 3.4. A class is defined through a set of instructions such as object construction, data storage and class member functions. The file containing the class definition takes a *.C extension, e.g., a class nc would be written in the file as nc.C. This file can be compiled independently of other code into a binary executable library file known as a shared object library with the *.so file extension, i.e., nc.so. When compiling a piece of code, say newApp.C (mypisoFoam.C in our case), that uses the nc class, nc.C need not be recompiled, rather newApp.C calls nc.so at runtime. This is known as dynamic linking.

As a means of checking for errors, the piece of code being compiled must know that the classes it uses and the operations they perform actually exist. Therefore, each class requires a class declaration, contained in a header file with a *.H file extension, e.g., nc.H, that includes the names of the class and its functions. This file is included at the beginning of any piece of code using the class, including the class declaration code itself. Any piece of *.C code can resource any number of classes and must begin with all the *.H files required to declare these classes. The classes in turn can resource other classes and begin with the relevant *.H files. By searching recursively down the class hierarchy, we can produce a complete list of header files for all the classes on which the top level *.C code ultimately depends; these *.H files are known as the dependencies. With a dependency list, a compiler can check whether the source files have been updated since their last compilation and selectively compile only those that need to be compiled.

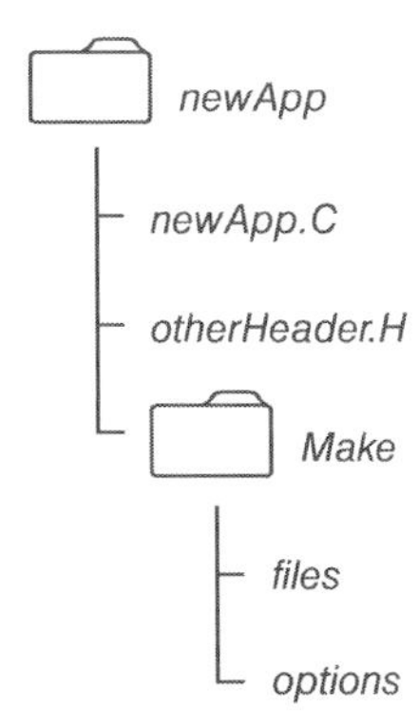

Figure 3.5 Directory structure for an application.

Header files are included in the code using # include statements, e.g., # include "otherHeader.H"; causes the compiler to suspend reading from the current file to read the file specified. Any self-contained piece of code can be put into a header file and included at the relevant location in the main code in order to improve code readability. For example, in most OpenFOAM applications the code for creating fields and reading field input data is included in a file createFields.H, which is called at the beginning of the code. In this way, header files are not solely used as class declarations. It is wmake that performs the task of maintaining file dependency lists amongst other functions.

OpenFOAM applications are organized using a standard convention that the source code of each application is placed in a directory whose name is that of the application. The top-level source file takes the application name with the *.C extension. For example, the source code for an application called newApp would reside is a directory newApp and the top-level file would be newApp.C as shown in Figure 3.5.

The compiler requires a list of *.C source files that must be compiled. The list must contain the main *.C file but also any other source files that are created for the specific application but are not included in a class library. For example, we may create a new class or add some new functionality to an existing class for a particular application. The full list of *.C source files must be included in the Make/files file. As might be expected, for many applications, the list only includes the name of the main *.C file, e.g., newApp.C in the case of our earlier example.

The Make/files file also includes a full path and name of the compiled executable, specified by the EXE = syntax. Standard convention stipulates the name is that of the application, i.e., newApp in our example. The OpenFOAM release offers two useful choices for path: standard release applications are stored in $FOAM_APPBIN; applications developed by the user are stored in $FOAM_USER_APPBIN.

As explained above, pisoFoam solver includes three dictionaries in its directory:

i Make directory, which includes two dictionaries (files, and options) for calling several header files and to address the compilation folder
ii (ii) createFields.H, which contains the required fields to be solved such as pressure and velocity fields. It was required to introduce concentration and temperature fields here as seen below.

```
Info<< "Reading field S\n" << endl;
volScalarField S
(
    IOobject
    (
        "S",
        runTime.timeName(),
        mesh,
        IOobject::MUST_READ,
        IOobject::AUTO_WRITE
    ),
    mesh
);

Info<< "Reading field T\n" << endl;
volScalarField T
(
    IOobject
    (
        "T",
        runTime.timeName(),
        mesh,
        IOobject::MUST_READ,
        IOobject::AUTO_WRITE
    ),
    mesh
```

iii pisoFoam.C, which is the main C++ code for the solver that includes the momentum equation and PISO loop. We need to add concentration and temperature equations here as well. These implementations are shown below.

```
fvScalarMatrix TEqn
(
    fvm::ddt(rho,T)
  + fvm::div(rhophi, T)
  - fvm::laplacian(kappaEff, T)
);

TEqn.relax();
TEqn.solve();


fvScalarMatrix SEqn
(
    fvm::ddt(rho,S)
  + fvm::div(rhophi, S)
  - fvm::laplacian(kappaEff, S)
);

SEqn.relax();
SEqn.solve();
}|
```

3.3.3 *Preparation of the case file*

The basic directory structure for an OpenFOAM case, which contains the minimum set of files required to run an application, is shown in Figure 3.6 and described as follows.

3.3.3.1 *The constant directory*

This contains a full description of the case mesh in a subdirectory polyMesh and files specifying physical properties for the application concerned, e.g., transportProperties.

3.3.3.2 *The system directory*

This sets parameters associated with the solution procedure itself. It contains at least the following three files: controlDict where run control parameters are set including start/end time, time step, and parameters for data output; fvSchemes where discretization

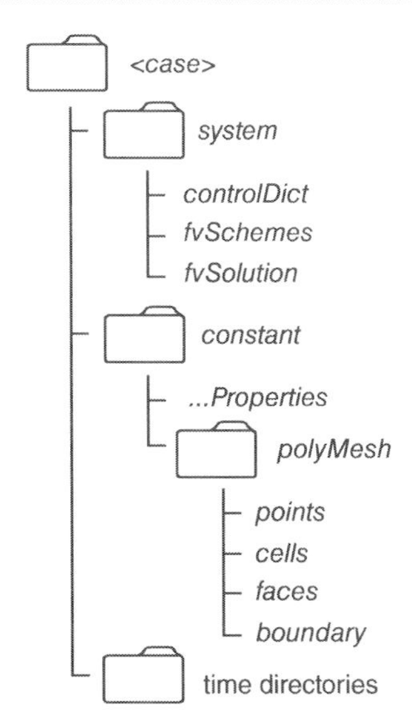

Figure 3.6 Case directory structure.

schemes used in the solution may be selected at run-time; and, fvSolution where the equation solvers, tolerances, and other algorithm controls are set for the run.

3.3.3.3 The "time" directories

These contain individual files of data for particular fields. The data can be either initial values and boundary conditions that the user must specify to define the problem, or results written to file by OpenFOAM. Note that the OpenFOAM fields must always be initialized, even when the solution does not strictly require it, as in steady-state problems. The name of each time directory is based on the simulated time at which the data is written.

The implementations and modifications, which have been done, are summarized as follows.

3.3.3.4 Constant directory

In this problem, constant directory includes polyMesh subdirectory where the mesh is built there using the blockMeshDict utility. Boundary conditions are also identified there. The domain is divided into a number of blocks. Each block should have eight vertices. All the vertices for all the blocks are sorted under the vertices dictionary. Each vertex has an x, y, and z coordinate value. The vertices are numbered in order starting from zero for the first vertex.

As mentioned, each block has eight vertices. These vertices are specified by their number according to their appearance in the vertices dictionary. The line for each block starts with the word "hex" meaning that the mesh will be hexahedral followed by the number of the eight vertices connecting the block. After specifying the vertices connecting one block comes the number of grid points in the three directions: x, y, and z.

The patches dictionary specifies the boundary patches in the mesh. Each boundary face should be connected by four vertices. First, the type of the boundary condition is specified for a patch and then the name of the patch. If a boundary patch is sharing many blocks, an entry for each piece of the patch in each block should be specified under the patch dictionary. The sample files for blockMeshDict and boundary are shown below, respectively.

```
convertToMeters 0.01;

vertices
(
        (0 0 0)//0
        (120 0 0)//1
        (0 0.6405 0)//2
        (120 0.6405 0)//3
        (0 1.5935 0)//4
        (120 1.5935 0)//5
        (0 50 0)//6
        (120 50 0)//7
        (0 0 19.5235)//8
        (120 0 19.5235)//9
        (0 0.6405 19.5235)//10
        (120 0.6405 19.5235)//11
        (0 1.5935 19.5235)//12
        (120 1.5935 19.5235)//13
        (0 50 19.5235)//14
        (120 50 19.5235)//15
        (0 0 20)//16
        (120 0 20)//17
        (0 0.6405 20)//18
        (120 0.6405 20)//19
        (0 1.5935 20)//20
        (120 1.5935 20)//21
        (0 50 20)//22
        (120 50 20)//23
```

```
7
(
    inlet
    {
        type            patch;
        nFaces          8;
        startFace       227375;
    }
    leftWall
    {
        type            wall;
        nFaces          1073;
        startFace       227383;
    }
    outlet
    {
        type            patch;
        nFaces          1081;
        startFace       228456;
    }
    lowerWall
    {
        type            wall;
        nFaces          1656;
        startFace       229537;
    }
    atmosphere
```

The transportProperties dictionary within the constant directory sets the transport model for the fluid to be Newtonian and then provides information about its viscosity and density. This is shown as follows.

```
transportModel  Newtonian;

nu              nu [ 0 2 -1 0 0 0 0 ] 1e-06;
mykappaEff      mykappaEff [ 1 -1 -1 0 0 0 0 ] 1e-03;

CrossPowerLawCoeffs
{
    nu0             nu0 [ 0 2 -1 0 0 0 0 ] 1e-06;
    nuInf           nuInf [ 0 2 -1 0 0 0 0 ] 1e-06;
    m               m [ 0 0 1 0 0 0 0 ] 1;
    n               n [ 0 0 0 0 0 0 0 ] 1;
}

BirdCarreauCoeffs
{
    nu0             nu0 [ 0 2 -1 0 0 0 0 ] 1e-06;
    nuInf           nuInf [ 0 2 -1 0 0 0 0 ] 1e-06;
    k               k [ 0 0 1 0 0 0 0 ] 0;
    n               n [ 0 0 0 0 0 0 0 ] 1;
}

// ************************************************************************* //
```

Table 3.2 RANS turbulence models for incompressible fluids within OpenFOAM

Model format in OpenFOAM	*Model name*
Laminar	Dummy turbulence model for laminar flow
kEpsilon	Standard high-Re k–ε model
kOmega	Standard high-Re k–ω model
kOmegaSST	k–ω-SST model
RNGkEpsilon	RNG k–ε model
NonlinearKEShih	Nonlinear Shih k–ε model
LienCubicKE	Lien cubic k–ε model
qZeta	q–ζ model
LaunderSharmaKE	Launder-Sharma low-Re k–ε model
LamBremhorstKE	Lam-Bremhorst low-Re k–ε model
LienCubicKELowRe	Lien cubic low-Re k–ε model
LienLeschzinerLowRe	Lien-Leschziner low-Re k–ε model
LRR	Launder-Reece-Rodi RSTM
LaunderGibsonRSTM	Launder-Gibson RSTM with wall-reflection terms
realizableKE	Realizable k–ε model
SpalartAllmaras	Spalart-Allmaras one-eqn mixing-length model

Moreover, in the turbulenceProperties the type of turbulence model is mentioned (RANS or LES) and in the other dictionary, the exact turbulence model should be mentioned regarding the turbulence type. OpenFOAM has many different turbulence models, which makes it a popular and strong application for simulations of turbulent flows. It contains various models for both incompressible and compressible fluids. These turbulence models are divided in two well-known categories: RANS (Reynolds Averaged Navier Stokes) and LES (Large Eddy Simulation).

The list of available RANS and LES models in OpenFOAM is presented in Tables 3.2 and 3.3.

3.3.3.5 System directory

The fvSchemes dictionary in the system directory sets the numerical schemes for terms, such as derivatives in equations, which appear in the applications being run. This section describes how to specify the schemes in the fvSchemes dictionary.

The terms that must typically be assigned a numerical scheme in fvSchemes range from derivatives, e.g., gradient ∇, and interpolations of values from one set of points to another. The aim in OpenFOAM is to offer an unrestricted choice to the user. For example, while linear interpolation is effective in many cases, OpenFOAM offers complete freedom to choose from a wide selection of interpolation schemes for all interpolation terms.

The derivative terms further exemplify this freedom of choice. The user first has a choice of discretization practice, where standard Gaussian finite volume integration is the common choice. Gaussian integration is based on summing values on cell faces,

Table 3.3 LES turbulence models for incompressible fluids within OpenFOAM

Model format in OpenFOAM	*Model name*
Smagorinsky	Smagorinsky model
Smagorinsky2	Smagorinsky model with 3-D filter
dynSmagorinsky	Dynamic Smagorinsky
homogenousDynSmagorinsky	Homogeneous dynamic Smagorinsky model
dynLagrangian	Lagrangian two equation eddy-viscosity model
scaleSimilarity	Scale similarity model
mixedSmagorinsky	Mixed Smagorinsky/scale similarity model
dynMixedSmagorinsky	Dynamic mixed Smagorinsky/scale similarity model
kOmegaSSTSAS	k−ω-SST scale adaptive simulation (SAS) model
oneEqEddy	k-equation eddy-viscosity model
dynOneEqEddy	Dynamic k-equation eddy-viscosity model
locDynOneEqEddy	Localized dynamic k-equation eddy-viscosity model
spectEddyVisc	Spectral eddy viscosity model
LRDDiffStress	LRR differential stress model
DeardorffDiffStress	Deardorff differential stress model
SpalartAllmaras	Spalart-Allmaras model
SpalartAllmarasDDES	Spalart-Allmaras delayed detached eddy simulation (DDES) model
SpalartAllmarasIDDES	Spalart-Allmaras improved DDES (IDDES) model

which must be interpolated from cell centers. The user again has a completely free choice of interpolation scheme, with certain schemes being specifically designed for particular derivative terms, especially the convection divergence $\nabla.$ terms.

The set of terms, for which numerical schemes must be specified, are subdivided within the fvSchemes dictionary into the categories listed in Table 3.4. Each keyword in Table 3.4 is the name of a subdictionary which contains terms of a particular type, e.g., gradSchemes contains all the gradient derivative terms such as grad(p) (which represents ∇p). Further examples can be seen in the extract from an fvSchemes dictionary as listed in Table 3.4.

Table 3.4 Main keywords used in fvSchemes

Keyword	*Category of mathematical terms*
interpolationSchemes	Point-to-point interpolations of values
snGradSchemes	Component of gradient normal to a cell face
gradSchemes	Gradient ∇
divSchemes	Divergence $\nabla.$
laplacianSchemes	Laplacian ∇^2
timeScheme	First and second time derivatives $\frac{\partial}{\partial t}$, $\frac{\partial^2}{\partial t^2}$
fluxRequired	Fields which require the generation of a flux

```
ddtSchemes
{
    default         CrankNicholson 0.5;
}

gradSchemes
{
    default         Gauss linear;
    grad(p)         Gauss linear;
    grad(U)         Gauss linear;

}

divSchemes
{
    default         none;

    div(phi,U)      Gauss cubic corrected;
    div(phi,S)      Gauss cubic corrected;
    div(phi,T)      Gauss cubic corrected;
    div(rhophi,S)       Gauss limitedLinear 1;
    div(rhophi,T)       Gauss limitedLinear 1;
    div(phi,k)      Gauss limitedLinear 1;
    div(phi,epsilon) Gauss limitedLinear 1;
    div(phi,R)      Gauss limitedLinear 1;
    div(R)          Gauss linear;
    div(phi,nuTilda) Gauss limitedLinear 1;
    div((nuEff*dev(T(grad(U))))) Gauss linear;
    div(nonlinearStress) Gauss linear;

}

laplacianSchemes
{
```

The equation solvers, tolerances, and algorithms are controlled by the fvSolution dictionary in the system directory. Below is an example set of entries from the fvSolution dictionary required for the mypisoFoam solver.

```
solvers
{
    p
    {
        solver          PCG;
        preconditioner  DIC;
        tolerance       1e-06;
        relTol          0.1;
    }

    pFinal
    {
        solver          PCG;
        preconditioner  DIC;
        tolerance       1e-06;
        relTol          0;
    }

    U
    {
        solver          PBiCG;
        preconditioner  DILU;
        tolerance       1e-05;
        relTol          0;
    }
//
    S
    {
        solver          PBiCG;
        preconditioner  DILU;
        tolerance       1e-05;
        relTol          0;
    }
    T
    {
        solver          PBiCG;
```

fvSolution contains a set of subdictionaries that are specific to the solver being run. However, there is a small set of standard subdictionaries that cover most of those used by the standard solvers. These subdictionaries include solvers, relaxationFactors, PISO, and SIMPLE algorithms (PISO has already been explained).

3.3.3.6 Linear solver control

The first subdictionary in our example, and one that appears in all solver applications, is solver. It specifies each linear-solver that is used for each discretized equation; it is

Table 3.5 Linear solvers in OpenFOAM

Solver	*Keyword*
Preconditioned (bi-)conjugate gradient	PCG/PBiCG[a]
Solver using a smoother	smoothSolver
Generalized geometric-algebraic multigrid	GAMG
Diagonal solver for explicit systems	diagonal

Note:
[a] PCG for symmetric matrices, PBiCG for asymmetric.

emphasized that the term linear-solver refers to the method of number-crunching used to solve the set of linear equations, as opposed to application solver, which describes the set of equations and algorithms used to solve a particular problem.

The syntax for each entry within solvers uses a keyword that is the word relating to the variable being solved in the particular equation. For example, pisoFoam solves equations for velocity U and pressure P, hence the entries for U and p. The keyword is followed by a dictionary containing the type of solver and the parameters that the solver uses. The solver is selected through the solver keyword from the choices in OpenFOAM, listed in Table 3.5. The parameters, including tolerance, and preconditioner are described in the following sections.

In the current study, for pressure field, PCG (Preconditioned Conjugate Gradient) is used for each discretized equation. PCG is a linear solver, the same as PBiCG (Preconditioned Bio Conjugate Gradient), which has been used for the other remaining fields, U, T, C, k, ε, and ω.

3.3.3.7 Solution tolerances

The sparse matrix solvers are iterative, i.e., they are based on reducing the equation residual over a succession of solutions. The residual is ostensibly a measure of the error in the solution so that the smaller it is, the more accurate the solution. More precisely, the residual is evaluated by substituting the current solution into the equation and taking the magnitude of the difference between the left and right hand sides; it is also normalized to make it independent of the scale of the problem being analyzed.

Before solving an equation for a particular field, the initial residual is evaluated based on the current values of the field. After each solver iteration, the residual is re-evaluated. The solver stops if any of the following conditions are reached:

- the residual falls below the solver tolerance, designated as tolerance.
- the ratio of current to initial residuals falls below the solver relative tolerance, relTol.
- the number of iterations exceeds a maximum number of iterations, maxIter.

3.3.3.8 Preconditioned conjugate gradient solvers

There are a range of options for preconditioning of matrices in the conjugate gradient solvers, represented by the preconditioner keyword in the solver dictionary. The preconditioners are listed in Table 3.6.

Table 3.6 Preconditioner options in OpenFOAM

Preconditioner	*Keyword*
Diagonal incomplete-Cholesky (symmetric)	DIC
Faster diagonal incomplete-Cholesky (DIC with caching)	FDIC
Diagonal incomplete-LU (asymmetric)	DILU
Diagonal	Diagonal
Geometric-algebraic multigrid	GAMG
No preconditioning	None

In numerical analysis and linear algebra, a preconditioner M of a matrix A is a matrix such that M-1A has a smaller condition number than A. Preconditioners are useful in iterative methods to solve linear system $Ax = b$ for x since the rate of convergence for most iterative solvers increases as the condition number of a matrix decreases as a result of preconditioning. In this example, DIC (Diagonal Incomplete Cholesky) preconditioner can be used for the pressure field. This is a simplified diagonal-based preconditioner for the symmetric matrices. However, DILU (Diagonal Incomplete LU) preconditioner is used for the other fields which mostly include asymmetric matrices to be solved.

3.3.3.9 Time control

The OpenFOAM solvers begin all runs by setting up a database. The database controls I/O and, since output of data is usually requested at intervals of time during the run, time is an inextricable part of the database. The controlDict dictionary sets input parameters essential for the creation of the database. A sample controlDict dictionary is shown in the following. Only the time control and writeInterval entries are truly compulsory.

```
application     pisoFoamIIII;
startFrom       startTime;
startTime       0;
stopAt          endTime;
endTime         90;
deltaT          0.001;
writeControl    adjustableRunTime;
writeInterval   1;
purgeWrite      0;
writeFormat     ascii;
writePrecision  6;
writeCompression off;
timeFormat      general;
timePrecision   6;
```

3.4 Postprocessing with ParaView

ParaView is an open-source platform based on the VTK (visualization toolkit) function library. VTK uses the Pipeline (pipeline) operating mechanism, which can process almost any type of data, and provides many corresponding classes to manipulate various types of data. The steps for VTK to visualize volume data are shown in Figure 3.7.

OpenFOAM can be installed in the Windows Subsystem for Linux (WSL), but a very troublesome problem is that WSL does not have a graphical interface, only a console window. During the post-processing process, users cannot directly call ParaView via the paraFoam command to start ParaView to post-process the calculation results of OpenFOAM. The essence of paraFoam is actually a script that calls ParaView. Users can rewrite this script and install ParaView under windows to post-process the calculation results of OpenFOAM. Once the results are written to the time directory, they can be viewed with ParaView. If a case is running with ParaView open, the output data on the time directory is not automatically loaded. At this point, clicking Apply in the Properties window will download the data to ParaView.

3.4.1 Isosurfaces and contour plots

When using ParaView post-processing calculation results to draw contours/surfaces, only scalars can be selected. If users want to draw contours/surfaces of vector components, such as the contours of velocity *ux*, what should they do?

Take the results from any simulation to make the contour of the velocity U_X. When contouring the profile, only the scalar p can be selected after contour, so if the velocity component can be extracted, it can be selected here. Delete contourl first, and with slicel selected, apply the Extract Component filter, set the properties, select u for the input array, x for the component, and name the output array *u-x*.

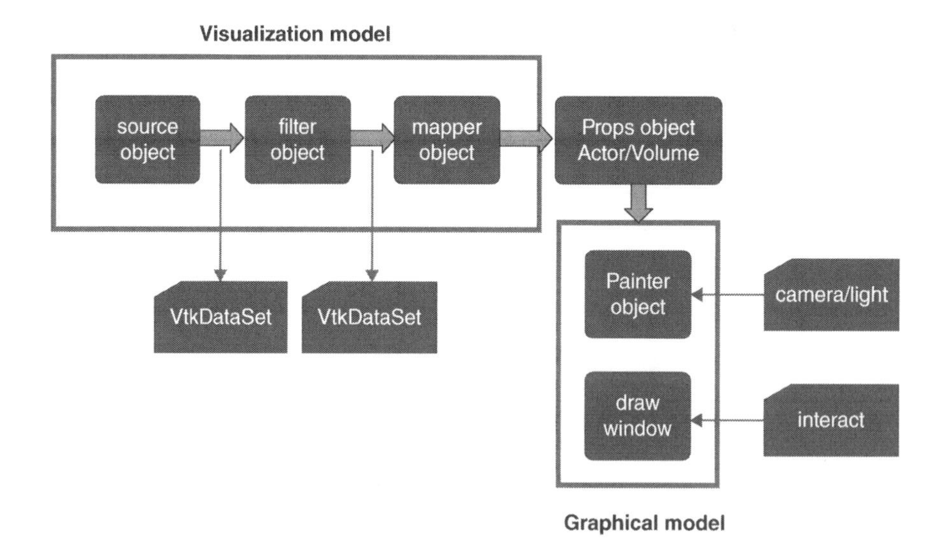

Figure 3.7 VTK visualization method.

In order to draw a simple pressure graph, in the Display panel, select the following: select the Color panel and rescale from color to data range via the menu. In the Style panel, choose Surface from the Representation menu; in the Color panel, choose p and Rescale to Data Range.

To view the solution for a moment, the user can change the current time to the target moment using the VCR Controls or the Current Time Controls located in the toolbar below the top menu of the ParaView window. Click the icon to get a continuous display, then click the icon to get a single pressure value in a cell; in this case, each cell will be represented by a single color with no grading.

If the model itself is 3D, the contours will be a series of 2D surfaces representing constant values, i.e., isosurfaces. The Contour's Properties panel includes a list of isosurfaces that can be edited, most conveniently using the New Range window, where the selected scale field is selected from the drop-down menu.

Color bars can be included by clicking the Toggle Color Legend Visibility button in the Active Variable Controls toolbar or by selecting Show Color Legend from the View menu.

To set a series of properties of the color bar, including text size, font selection, and numbering format of the ticks, the user should click the Edit Color Map button in the Active Variable Controls toolbar or the Color panel in the Display window. The color bar can be dragged with the mouse. Position in the image window.

If users use a newer version of ParaView, they will see that ParaView uses a blue to white to red color scale by default instead of the more common blue to green to red (rainbow)

To change the color stop, the user can do the following: select a selection preset in the color stop editor, then choose a blue to red rainbow. If the user wants to use this color bar all the time, after clicking the "OK" confirmation button, the user can click the "Set Default" button. At this point, the color of the geometric surface is changed by the user, and the result of the change can be seen by rotating the image.

3.4.2 Vector plots

It is best to delete other created modules before the flow velocity vector, there are two ways: to delete completely, highlight the relevant module in the Pipeline Browser, and click Delete in the respective Properties panel; and by toggling the eye button of the relevant module in the Pipeline Browser to disable.

1 At the beginning, ensure that the OpenFOAM file is active, i.e., the small eye icon in front of the file is bright, then select filter->Alphabetical (the last option)->cell centers, and then click Apply.
2 At this time, the small eyes in front of the cell center are active. Then perform the following operations: filter——>Alphabetical——>Glyph.
3 Set: In the orient option in the Object Inspector, because it is a Vector, select OFF in scale Mode, and fill in an appropriate value in the set scale factor, then try to observe the effect.
4 Coloring, if there is no color, you can set it in display
5 Users can try to change the parameters in ARROW to make the picture more visually appealing.

The speed can be represented by arrows in ParaView. To make a simple vector diagram, do the following: open the prepared file, find the generated file, the user can select all grid elements and all variables to apply, the default display is its pressure information, the user can enter u to view its speed information, for quality problems users can choose Glyph, which can make a vector plot, which is to use arrows to represent the speed and size of the water flow, click Glyph, click Apply, and click play to view the animation. ParaView's default arrow size changes with pressure, but users may want more arrow sizes with velocity u. Users can select u in the scale array to view the speed information, and click Apply after conversion.

The user can choose to play the animation from the beginning. Click the scale factor to change the arrow size. When users do not want to change the size of the arrow, but only want to know the direction of the flow velocity, they can select the unscaled array from the scale array, click Apply, and click Play to view the animation, which will clearly show the direction of the flow velocity.

To add an arrow, select Stream Tracer, add Glyph (upper left corner) on it, change the properties of Glyph: Glyph Type to Cone (cone); (arrow shape is cone) Orientation Array to u (speed); Scale Array to For u; (set the arrow size according to the speed) Scale Factor reset (reset button) (adjust the overall arrow size). Colored by temperature, with Glyph still selected in the pipeline window, change the coloring to Temp. This way, the size of the arrow represents the magnitude of the velocity, and the color of the arrow represents the temperature.

Note that although the glyph is represented as flowing through the wall, its size is zero because ParaView chooses to orient the glyph in the x-direction when the glyph shrinks and the velocity size is 0.

3.4.3 Streamline plots

The user should disable modules, such as those for the vector illustration above, before continuing post-processing in ParaView.

Now, in order to draw velocity streamlines, keep the cavity.OpenFOAM module highlighted in the Pipeline Browser to generate high quality streamline images: first, the user should select Stream Tracer from the Filter menu, then click Apply, and second, set according to the specific model Parameters, specify the Seed points (origin) along the Line Source, which runs perpendicular to the geometric center. Click Apply to generate the trajectory, and then select the Tube from the Filter to generate a high-quality streamline diagram. Then click Apply.

1 Obtain a 3D streamline diagram
   Proceed as follows:
      After opening the chosen model, select Filters -> Stream Tracer
      Without changing any parameters, some streamline diagrams in 3D can be obtained.
   Modify the Seeds setting parameters:
   a Using the point Source setting:
      1 Select point Source
      2 Determine the position/coordinates of the point
      3 Determine the number of points n

4 Determine the distribution radius of points, and randomly distribute n points in the spherical space with this value as the radius

b Use the line Source setting:

1 Select high resolution line Source
2 Determine the positions/coordinates of the two ends of the line
3 Determine the resolution n

The number of points thus obtained is $n + 1$ (including both endpoints)

Tips:

The user can also click and hold the point to be moved with the left mouse button. While pressing the left mouse button, moving the mouse can also change the position of the point, and releasing the left mouse button can determine the position of the point.

Usually, the default is 3D display. In this 3D mode, the user can scroll the middle button of the mouse to zoom. Press the middle button and drag the mouse to pan the model. If you hold the left button of the mouse and drag, you can rotate the graphic. There are also a series of tabs; the user can click these buttons to lock the viewing angle in a specific orientation of some coordinate values. For example, to observe the model from the positive direction of the y-axis, click the button below.

However, sometimes the results of some examples are two-dimensional, and the corresponding velocity is also two-dimensional, with only two components in the x and y directions. After loading the results in ParaView, if the user wants to draw the streamline, they will find that the streamline drawing button Stream Tracer is grayed out and the function is unavailable. In this case, how should the user draw the streamline of the 2D velocity vector?

First of all, it needs to be clear that the reason why Stream TracerQ is unavailable is that ParaView can only recognize 3D vectors, and many functions only work on 3D vectors. Therefore, 2D vectors need to increase the component in the z direction. The Calculator filter can be used to create new 3D velocity vectors.

2 Cross-sectional streamline diagram

The Stream Tracer of ParaView cannot directly make a streamline on a section. For example, for the OpenFOAM example, even if it is a two-dimensional example, after intercepting a surface, the streamline cannot be obtained with Stream Tracer. There is a solution. The following describes how to obtain streamlines on the section through a combination of a series of filters. Taking the pitzdaily example, the steps are as follows:

a Make a section (slice)

b Use the Surface Vector filter on the section; the purpose of this is to project the velocity vector onto the plane.

c Use the Mask Points filter on the obtained SurfaceVector. The purpose of this is to generate a series of reference points. When drawing the streamline in the future, the position and density of the streamline will be determined by the position of these reference points.

The On Ratio parameter controls the density of the points; Maximum number of points controls the total number of points; Random Sampling, enables the random point mode, if it is a nonrandom mode, the points will be selected according to the coordinates from small to large. Assuming that On Ratio = 2560, there are a total of 1000 points, but if the Maximum number of points is set to 100, then only the first 100 points with the smallest coordinates will be taken, and if the random mode is turned on, the distribution of points is basically uniform fill for this flow area. Generate Vertices, select whether to display reference points; if enabled, a dot matrix will be displayed.

d Select Stream Tracer with Custom Source, Input, and Seed Source in Filter The streamline diagram of the section is obtained. The density of the streamline can be controlled by the number of Mask Points.

3.4.4 Two ways for ParaView to create animation

First save the image sequence in ParaView. Click file>save Animation to save image sequences in tif, png, jpg, and other formats. The number of frames to be saved can be specified; the default is one frame per timestep. Note: If there are many pictures, it is best to create a new folder. Secondly, there are two ways to convert the picture sequence to video or GIF.

The first way: use the third-party post-processing tool with OpenFOAM, type in the terminal under the example folder. The specific usage of foamCreateVideo can be viewed with foamCreateVideo -help.

The second way: use the convert/usr/bin/convert -quality 90% frames*.png movie. gif in the Ubuntu system. Obviously, -quality 90% means the video quality, where frames are the image sequence name, and * means any name that follows, such as frames.000.png, movie is the name of the generated gif.

References

Gildeh, H.K., Mohammadian, A., and Nistor, I. Inclined dense effluent discharge modelling in shallow waters. *J. Environ. Fluid Mech.* 2021, 21, 955–998. doi:10.1007/s10652-021-09805-6.

Issa, R.I. Solution of the implicitly discretized fluid flow equations by operator-splitting. *J. Comput. Phys.* 1985, 62, 40–65.

Jasak, H. *Error analysis and estimation for the finite volume method with applications to fluid flows.* Ph.D. Thesis, Imperial College of Science, Technology and Medicine, 1996.

Jasak, H., Weller, H., and Nordin, N. In cylinder CFD simulation using a C++ object-oriented toolkit. *SAE Technical Papers*, 2004.

Juretic, F. *Error analysis in finite volume.* Ph.D. thesis, Imperial College of Science, Technology and Medicine, 2004.

Millero, F.J. and Poisson, A. International one-atmosphere equation of state of sea water. *J. Deep-Sea Res.* 1981, 28A(6), 625–629.

Chapter 4

Applications

4.1 Review of past numerical studies in the field[1]

4.1.1 Discharge through inclined dense jets

An inclined jet is discharged at an angle of 0–90° from the horizontal and may be positively or negatively buoyant. Most of the inclined jets used in engineering applications are negatively buoyant jets, which are also known as inclined dense jets. Most high-density discharge outfalls are designed as inclined jets due to their proven near-field dilution rate and efficiency.

Many new computational fluid dynamics (CFD) techniques were developed in the 1970s, but detailed and reliable experimental data were difficult to come by. In the 1990s, particle image velocimetry (PIV) and laser-induced fluorescence (LIF) experimental techniques provided more reliable data, and these studies became benchmarks for CFD model validation. Bemporad (1994) and others performed some numerical studies on jet mixing in the 1990s, but Vafeiadou et al. (2005) were the first to fully apply CFD tools to estimate the effluent discharge jet flow behavior. Vafeiadou simulated dense effluent discharges at 45°, 60°, 70°, 80°, and 90° inclinations, by using ANSYS CFX with a mesh of 400,000 elements and adopting the k-ω SST (shear stress transport) turbulence model. Comparing Vafeiadou's results with experimental data by Bloomfield and Kerr (2002) and Roberts et al. (1997), it appeared that their numerical results agreed, supporting the further use of such numerical models to estimate the characteristics of dense effluent discharges. Vafeiadou's study pioneered the use of CFD for the mixing of effluent discharges into water bodies.

Oliver et al. (2008) also used ANSYS CFX and applied the standard k-ε turbulence model for the RANS equations closure. Experimental data of dense effluent discharges were simulated by the k-ε turbulence model using both a standard and adjusted turbulent Schmidt number of $S_{ct} = 0.9$. Both of these simulations with the k-ε turbulence model were more accurate than the integral models and the analytical solutions. In the experimental studies, the buoyancy-induced instabilities were observed on the lower (inner) half of the jet. However, in the study of Oliver et al. (2008), the k-ε simulations (both standard and calibrated) underpredicted the spread of the jet and the integrated centerline dilution at the top rise of the jet, because they overestimated the influence of the stabilizing density gradients. This

1 This section is updated and based on Mohammadian et al. (2020).

DOI: 10.1201/9781003181811-4

study only achieved qualitative comparisons and did not arrive at any conclusions regarding which model performed the best.

Gildeh et al. (2015) used OpenFOAM to explore the numerical modeling of 30° and 45° inclined dense jets in calm ambient water conditions. OpenFOAM has a solid structure of various solvers, including a wide range of turbulence models that can be used in jet mixing and dispersion analysis. This study focused on how the choice of the turbulence model can affect the jet trajectory and predicted dilution, and the importance of turbulence closure in the Navier-Stokes equations. They applied five RANS turbulence models to investigate the accuracy of the CFD predictions: the Launder-Reece-Rodi (LRR) Reynolds stress model (RSM), the Launder-Gibson RSM, the RNG k-ε linear eddy viscosity model, the realizable k-ε linear eddy viscosity model, and a nonlinear k-ε eddy viscosity model. A summary of these turbulence models is discussed in Gildeh (2013).

Gildeh et al. (2015) discretized the temporal term with a first-order implicit Euler scheme. The standard finite volume method with a Gaussian integration was used to discretize the advection-diffusion terms. The preconditional conjugate gradient (PCG) scheme was used to solve the pressure field, and the preconditioned biconjugate gradient (PBiCG) scheme was used for the other fields: U, T, C, k, ε, and ω. This study focused on analyzing the inclined dense discharge, including the jet trajectories, jet terminal rise height, jet centerline peak, jet centerline, and jet horizontal return point (i.e., impact point). The modeled values for velocity and concentration profiles were compared to experimental data. The effects of the increased jet velocities of the high Froude numbers observed in field applications were investigated via a sensitivity analysis. Gildeh et al. (2015) found that of the turbulence models tested in their study, the realizable k-ε and LRR models were more accurate than the other three tested. Tables 4.1 and 4.2 (after Gildeh et al. 2015) summarize the results for these two models (only 30° jets are presented for the sake of brevity).

The geometry, concentration, and velocity fields in the near field of effluent discharges can be characterized when the jet flow pattern is understood. Figure 4.1 presents the normalized concentration maps (C/C_0, where C is the computed concentration and C_0 is the discharge concentration) and the dilution isolines for inclinations of 30° and 45°; realizable k-ε and LRR turbulence models were used for each case.

Effluent discharge trajectory is an important factor in the design of ocean outfall systems. The discharge trajectory basically identifies the flow path that a jet would travel through until it impacts the bed. A dilution equation (Equation (4.1); C_a is ambient concentration) was used to calculate the dilution values for plotting dilution in Figure 4.1. The discharge trajectory and flow growth are also illustrated in Figure 4.1 as the jets travel downstream with clear differences between the two turbulence models. LRR is an anisotropic turbulence model and performs more reliably for calculating the shear forces on jet edges, and provides more realistic predictions. Tables 4.1 and 4.2 summarize the discharge trajectory and flow growth compared to other turbulence models.

$$S = C_0\text{-}C_a / C\text{-}C_a \tag{4.1}$$

The jet centerline often follows the maximum cross-sectional velocity or concentration along the flow path perpendicular to the discharge trajectory. The best way to extract the centerline is to start at the nozzle and create a velocity vector map. Shao

Table 4.1 Comparison of numerical and experimental coefficients for 30° inclined jets

		Gildeh et al. (2014)		Shao and Law (2010)		Kikkert et al. (2007)					
Parameter	*Proportionality coefficient*	*Realizable k-ε*	*LRR*	*$0.10 \leq y_0/L_m \leq 0.15$*	*$y_0/L_m > 0.15$*	*LA data*	*LIF data*	*Theory*	*Nemlioglu and Roberts (2006)*	*Cipollina et al. (2005)*	*Zeitoun et al. (1970)*
Terminal rise height	$\frac{Y_t}{L_M}$	1.13	1.13	1.13	-	1.14	1.34	1.08	1.49	1.15	1.22
Horizontal location of return point	$\frac{X_r}{L_M}$	3.40	3.34	3.06	3.19	3.34	3.66	3.14	3.51	3.22	3.70
Return point dilution	$\frac{S_r}{Fr}$	1.27	1.31	1.18	1.45	-	-	-	1.90	-	-
Vertical location of centerline peak	$\frac{Y_m}{L_M}$	0.71	0.69	0.70	-	0.59	0.70	0.66	-	0.84	-
Horizontal location of centerline peak	$\frac{X_m}{L_M}$	2.05	1.97	1.81	1.64	1.86	1.97	1.81	-	2.07	-
Centerline peak dilution	$\frac{S_m}{Fr}$	0.65	0.63	0.62	0.66	-	-	-	-	-	0.36

Source: Gildeh et al. (2015).

Note:
LIF: Laser-Induced Fluorescence, LRR: Launder-Reece-Rodi.

Table 4.2 Performance evaluation of two turbulence models (realizable k-ε and LRR models) for 30° inclined jet

		(Gildeh et al., 2015)			*Absolute difference (%)*	
Parameter	*Proportionality coefficient*	*Realizable k-ε*	*LRR*	*Average of experiments*	*Realizable k-ε*	*LRR*
Terminal rise height	$\frac{Y_t}{L_M}$	1.13	1.13	1.22	8.09	8.09
Horizontal location of return point	$\frac{X_r}{L_M}$	3.40	3.34	3.37	1.04	0.75
Return point dilution	$\frac{S_r}{Fr}$	1.27	1.31	1.51	18.90	15.27
Vertical location of centerline peak	$\frac{Y_m}{L_M}$	0.71	0.69	0.70	1.91	0.97
Horizontal location of centerline peak	$\frac{X_m}{L_M}$	2.05	1.97	1.86	10.22	5.91
Centerline peak dilution	$\frac{S_m}{Fr}$	0.65	0.63	0.55	18.90	15.24

Source: Gildeh et al. (2015).

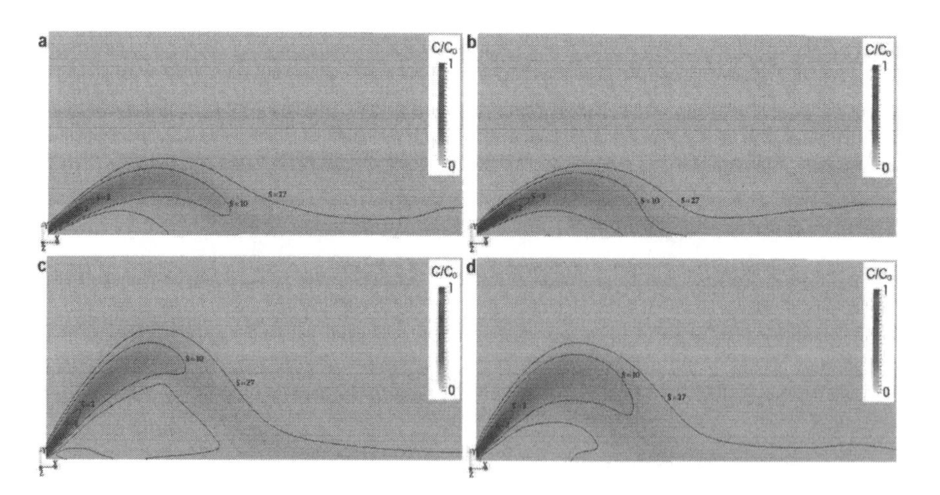

Figure 4.1 Mixing regimes for 30° and 45° inclined dense jets. (a) 30°, realizable k-ε; (b) 30°, LRR; (c) 45°, realizable k-ε; and (d) 45°, LRR (Gildeh et al., 2016).

(2009) noted that the maximum concentration and velocity profiles almost coincide, but the maximum concentration profile usually decreases more quickly than the velocity. The momentum and the buoyancy-induced instabilities also affect the trajectories of inclined dense discharges. While the discharge rises close to the nozzle, negative buoyancy forces affect the upward momentum and decrease it until they dominate the concentration transport (somewhere after the discharge rise peak) and the maximum concentration profile sinks down toward the bed faster than the maximum velocity profile. The centerline results in (Gildeh et al., 2015) were extracted following the maximum cross-sectional velocity similar to (Shao, 2009).

Gildeh et al. (2016) used the buoyancy modified standard *k-e* model and the two best performing models in Gildeh et al. (2015) (realizable *k-e* and LRR) for inclined dense jets. They modified the standard k-ε turbulence scheme to include the standard Boussinesq gradient diffusion hypothesis (SGDH) and the general gradient diffusion hypothesis (GGDH). The governing equations for the k-ε turbulence model that include the buoyancy term are as follows:

$$\frac{\partial k}{\partial t}+\mu_i\left(\frac{\partial k}{\partial x_i}\right)=\frac{\partial}{\partial x_i}\left(\frac{v_t\,\partial k}{\sigma_k\,\partial x_i}\right)+P+G-\varepsilon \tag{4.2}$$

$$\frac{\partial \varepsilon}{\partial t}+\mu_i\left(\frac{\partial \varepsilon}{\partial x_i}\right)=\frac{\partial}{\partial x_i}\left(\frac{v_t\,\partial \varepsilon}{\sigma_\varepsilon\,\partial x_i}\right)+C_{1\varepsilon}\frac{\varepsilon}{k}(1-C_{3\varepsilon})(P+G)+C_{2\varepsilon}\frac{\varepsilon^2}{k} \tag{4.3}$$

where the shear production term P and the buoyancy production term G in the k-ε turbulence model in Equation (4.2) are calculated as follows:

$$P=v_t\left(\frac{\partial u_i}{\partial x_i}+\frac{\partial u_j}{\partial x_i}\right)\frac{\partial u_i}{\partial x_j} \tag{4.4}$$

$$G=\frac{\overline{\rho' u_i'}}{\bar{\rho}^2}\left(\frac{\partial P}{\partial x_i}+\rho g_i\right) \tag{4.5}$$

The way the term $\overline{\rho' u_i'}$ in Equation (4.5) is solved defines the SGDH and GGDH methods. The corresponding equations for SGDH and GGDH are as follows, respectively.

$$\overline{\rho' u_i'}=\frac{-v_t}{\mathrm{Pr}_t}\frac{\partial \bar{\rho}}{\partial x_j} \tag{4.6}$$

$$\overline{\rho' u_i'}=\frac{-3}{2}\frac{C_\mu}{\mathrm{Pr}_t}\frac{k}{\varepsilon}\left(\overline{u_i' u_j'}\frac{\partial P}{\partial x_i}\right)=\frac{-3}{2}\frac{v_t}{\mathrm{Pr}_t\,\rho^2 k}\left(\overline{u_i' u_j'}\frac{\partial \bar{\rho}}{\partial x_i}\right) \tag{4.7}$$

The modified SGDH and GGDH k-ϵ turbulence models were used in OpenFOAM to explore the influence of the buoyancy term, using turbulence coefficients $C_\mu = 0.09$, $C_{1\varepsilon} = 1.44$, $C_{2\varepsilon} = 1.92$, and a calibrated coefficient, $C_{3\varepsilon}$ for sensitivity tests with $C_{3\varepsilon} = 0.9$, 0.6, and 0.4.

Figure 4.2 shows numerical and experimental results for a 45° inclined dense jet on the central plane. The standard k-ε results are relatively similar to experimental data,

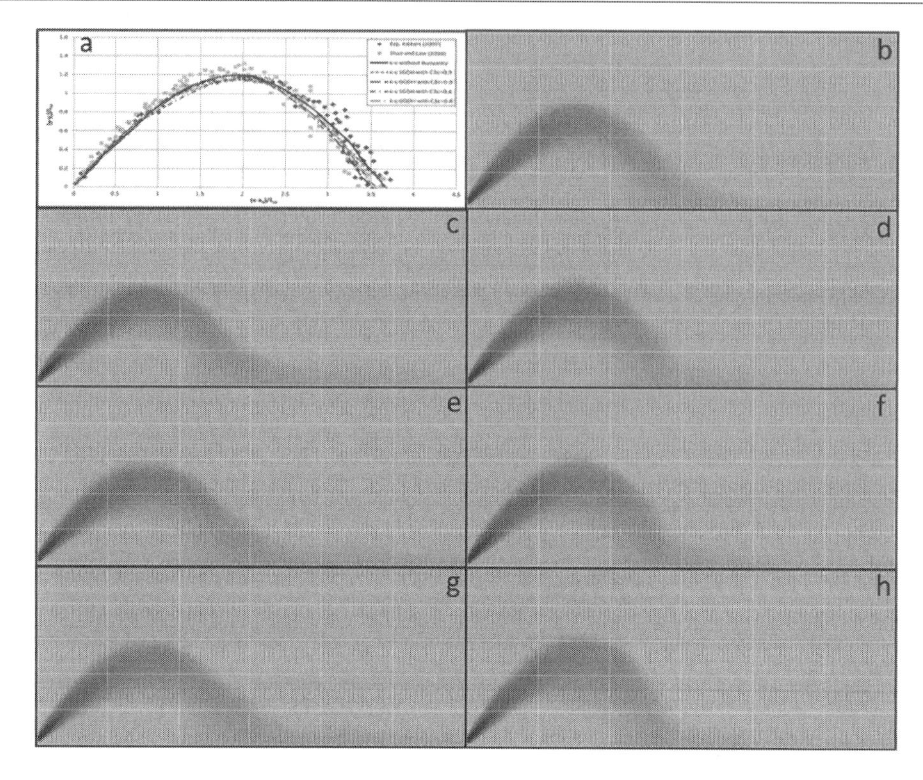

Figure 4.2 Overall discharge trajectory for 45° inclined dense jets. (a) Centerline comparison (modeled vs. experiments); (b) standard *k*-ε; (c) modified *k*-ε with SGDH, $C_{3\varepsilon} = 0.9$; (d) modified *k*-ε with GGDH, $C_{3\varepsilon} = 0.9$; (e) modified *k*-ε with SGDH, $C_{3\varepsilon} = 0.6$; (f) modified *k*-ε with GGDH, $C_{3\varepsilon} = 0.6$; (g) modified *k*-ε with SGDH, $C_{3\varepsilon} = 0.4$; and (h) modified *k*-ε with GGDH, $C_{3\varepsilon} = 0.4$ (Gildeh et al., 2016).

both with and without modifications (Figure 4.2a). Adding the buoyancy term to the turbulence model causes the jet to spread more along the inner (lower) half of the jet, where stronger buoyancy-induced forces would be experienced. When the buoyancy effects are set to be stronger (i.e., with a smaller $C_{3\varepsilon}$, the discharge growth rate is larger, and the numerical results align more closely to the experimental data (not presented in this chapter). The SGDH and GGDH methods showed similar results when the computed and experimental discharge trajectories were compared. This could be attributed to the small difference in density between the discharge and the receiving water (smaller than 1%). The buoyancy effect on the effluent discharge mixing characteristics should be investigated further, especially for larger density differences between the discharge and ambient water (larger than 1% in this study) as well as $C_{3\varepsilon}$ calibration for mixing applications.

Gildeh et al. (2016) further studied the concentration and velocity effects on the dispersion properties and the discharge growth rate for the 30° and 45° inclinations (Figure 4.3). The figure presents the normalized cross-sectional concentrations for the 30° and 45° discharges using the realizable k-ε and LRR turbulence models. The realizable k-ε model predicts a narrower jet width than the LRR model does. The jets predicted by Linear Eddy Viscosity Models (LEVMs) are usually thinner than

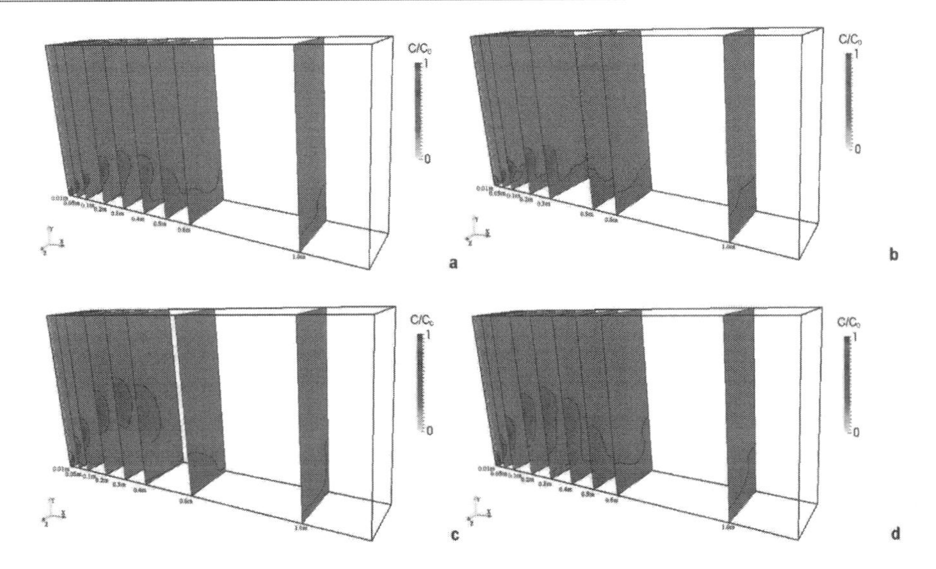

Figure 4.3 Discharge growth width at various cross-sections: the contour lines represent $S = C_0/C = 1$. (a) 30°, realizable *k*-ε; (b) 30°, LRR; (c) 45°, realizable *k*-ε; and (d) 45°, LRR (Gildeh et al., 2016).

the ones predicted using the RSMs, due to the assumption of isotropic turbulence in the LEVMs. The improved behavior of the LRR model could be due to stress anisotropy.

LEVMs are less expensive numerical models than the RSMs. There are reasons that justify extra terms in the RSMs over the two-equation models such as *k*-ε for the mixing application. To begin with, Reynolds stresses are affected by the pressure strain term through both the turbulent fluctuations (which is related to the gradient in mean velocity) and to the turbulence-turbulence interaction (effects of eddies in the fluctuation of pressure), which do not relate directly to the mean flow changes. The first process is called rapid pressure, and the second is known as slow pressure.

This affects the energy redistribution between stress components, which is important in estimating the degree of the stresses' anisotropy. Inclined dense discharges disperse according to the interaction of eddies. Therefore, both the rapid pressure and slow pressure terms may be important. It is possible that the reason the LRR model works well is that it includes both the slow and rapid pressure terms.

OpenFOAM was used to simulate 45° inclined discharges using the LES method and the Smagorinsky and Dynamic Smagorinsky subgrid scale (SGS) by Zhang et al. (2016). This resulted in numerical predictions including geometrical characteristics, the jet trajectory, jet spread, and eddy structures. Eddy sizes in the LES method are based on the grid spacing. In LES models, large eddies are simulated directly by computing the Navier-Stokes equations, but small eddies are simulated using, for instance, Boussinesq hypothesis assumptions. The simulations were run with the twoLiquidMixingFoam solver in OpenFOAM. Most of the parameters in Table 4.1 were used to compare the study results of Zhang et al. (2016) with other experimental and numerical studies. Eddy structures and turbulence characteristics were computed and compared to the experimental data of Oliver et al. (2013), Papakonstantis et al. (2011a), and

Papakonstantis et al. (2011b). The coherent structure of the inclined dense discharge plays an important role in the development of flow in the jet and can help elucidate the mixing properties and the intensity of turbulence.

LES seemed to underestimate the turbulence intensity in the area of buoyancy-induced instabilities (i.e., lower half of the jet), as reported by Zhang et al. (2016). For turbulent flows, the effect of the small scales on the simulated resolved scales are accounted for with an SGS approach. To complete the closure, for the Smagorinsky model, the parameter (C_S) in the eddy viscosity, $v_t = \rho(C_s\Delta)^2 S_{ij}$ where Δ is the LES filter size and S_{ij} is the strain rate tensor, needs to be specified. The accuracy in choosing this parameter is directly related to the kinetic energy dissipation.

In 1991, Germano et al. (1991) made an important addition to turbulence theory and modeling when they proposed a model for the dynamic determination of C_S. Rather than obtaining the minimal coefficient from predetermined expressions, the model analyzed large-scale turbulences during the numerical simulation to deduce them. This dynamic model approach was proven to be more applicable in the cases with complex interactions.

Robinson et al. (2015) predicted the mixing and dispersion of inclined dense discharges in stagnant surroundings by developing a three-dimensional CFD model that uses the CFD tool named Fluidity (Piggott et al., 2008). Fluidity uses a range of control volumes and finite element discretization to solve the 3D Navier-Stokes equations using unstructured mesh; it is capable of including Coriolis effects, density variations, turbulence models, tidal forcing, and associated buoyancy forces. Fluidity includes an anisotropic adaptive mesh capability that controls the solution accuracy locally throughout the model domain. The standard k-ϵ and V-LES (very-large eddy simulation) models in the dense jet simulations were used to close the Navier-Stokes equations. Although they only studied the 60° jet for the centerline rise height, impact point, and its dilution, and though their study was preliminary and not detailed in terms of model validation and comparisons, it seems that their model underpredicted the terminal rise height, the distance from the nozzle to the impact point, and the minimum dilution at impact point compared to the experimental data. This could be due to minimal water entrainment toward the jet centerline; therefore, the simulated jet is not as diluted as it should be. This could be attributed to the mesh used in their preliminary study.

Zhang et al. (2017) built upon their previous study (Zhang et al., 2016) to include the tail of the return point density current analysis with the LES model in OpenFOAM. The standard k-ε RANS turbulence model was run to compare the results with the LES model, and the 45° and 60° inclinations were selected. The jet spreading layer thickness was defined as the lateral distance where the concentration would reach 5% of the maximum concentration at the jet centerline. The 60° jet spreading layer results were compared to those from experiments by (Nemlioglu and Roberts, 2006). Similar to the study of (Zhang et al., 2016), the eddy structures and concentration profiles were plotted along the jet trajectory, and the results showed that LES was well able to predict the return point location and the coordinates of the centerline peak. However, their model underestimated the dilution at the return point by approximately 20% compared to data collected by previous experiments. This could be due to the mesh resolution that was adopted in this study, as they argued. Compared to the experimental data, LES was able to reproduce the concentration density in the density current region, while k-ε results could not. It was observed that for both geometrical and flow/dilution characteristics, k-ε results were generally lower in values compared to the LES results.

Jiang and Law (2018) built on the work of Zhang et al. (2016) and Zhang et al. (2017) by studying the effect of swirls at the outfall nozzles. Experimentally, they found that the addition of swirls at the nozzles lowered the terminal rise height of inclined dense discharges and increased the dilution rate at the return point. This could potentially affect the outfall design for coastal waters. Their numerical study is the first of its kind for the CFD modeling of swirls in effluent discharges. Similar to Zhang et al. (2017), two approaches were taken into consideration for this study: (i) LES using the dynamic Smagorinsky subgrid approach and (ii) RANS using the standard k-ε turbulence model. In the case with swirls at the nozzle, RANS exhibited better performance than LES when the results were compared to the experimental data, which could be due to the reduction in turbulence anisotropy and more axisymmetric distribution of turbulence. On the other hand, in the case of no swirls at the nozzle, the performance of LES was superior, as it could account for the anisotropy in a more accurate manner. When the swirls number (maximum tangential velocity divided by the discharge velocity at the nozzle) was larger than 0.33, it was observed that both RANS and LES models underestimated the dilution of discharge.

Yan et al. (2020) used a numerical model with RNG k-ε turbulent closure to model inclined plane jets in a linear layered environment, and proposed a new empirical function of dimensional analysis-based characteristic parameters for important online parameters for rapid estimation of inclined plane jets in a linearly layered environment. By comparing the numerical simulation results with experimental measurements, it showed that the numerical model can provide satisfactory predictions for the variables of interest.

Ramezani et al. (2021) numerically investigated the mixing and geometric properties of 30° and 45° inclined dense jets as they exit close to the bed. Two series of numerical simulations were developed to investigate the effect of close proximity to the bed on the behavior of dense jets. Numerical results were presented in the form of comparison plots and compared with previous work. The comparison showed that the proximity to the bed has little noticeable effect on the flow behavior of the 45° jet, but some changes in the geometry and mixing properties of the 30° jet are observed. The numerical model can better predict the geometric shape of the model and the dense jet characteristics are in good agreement with previous experimental studies. However, dilution predictions are conservative for both free and boundary-influenced jets.

Gildeh et al. (2021) presented the numerical results for inclined dense effluent discharges into a shallow receiving water body with surface impingement for the first time. Two jet discharge angles relative to the horizontal (30° and 45° were selected for this study. Five RANS turbulence models were examined in their study: realizable k-ε and k-ω SST models (known as two-equation turbulence models), v2f (four equations to model anisotropic behavior), and LRR and SSG turbulence models (known as RSMs – six equations to model anisotropic behavior). Three mixing regimes introduced in Jiang et al. (2014) were reproduced numerically for both discharge angles applying various turbulence models: full submergence, plume contact, and centerline impingement regimes (i.e., FSR, PCR, and CIR). Key geometrical and dilution properties of these jets at surface contact (X_s, S_s) and jet centerline return point (X_r, S_r) were compared to those available from experiments. Normalization parameter that was selected for jets in shallow waters was H/D (water depth above discharge point over nozzle diameter).

It was found that surface attachment increases the return point length from the nozzle and that surface dilution decreases from FSR toward CIR. Among turbulence models tested, RSMs (LRR and SSG) predicted the effluent discharge kinematic and dilution properties better compared to two- and four-equations models. This was mainly attributed to the anisotropic nature of the effluent discharge problem studied herein and that these models are better capable to account for anisotropies.

4.1.1.1 Discussion of differences in RANS and LES models for effluent mixing problems

RANS and LES models are very popular models in the modeling of effluent discharges in ambient waters, as reviewed in the studies above. Navier-Stokes equations were previously described in Chapter 2. Reynolds decomposition involves splitting any instantaneous quantity into mean and fluctuating components by time averaging for both velocity and momentum equations. RANS equations resemble basic governing equations, except for the turbulent momentum and density fluxes, which result from the Reynolds decomposition and averaging. Six unknown terms of the Reynolds stress tensor and three unknown terms of the density flux term imply that the number of unknowns is larger than the available equations. Therefore, this leads to an undetermined system of equations commonly referred to as the closure problem. To resolve this severe shortcoming, several hypotheses and methods have been prescribed. The turbulent viscosity and gradient-diffusion hypotheses are the most widely used concepts to deal with the closure problem. The turbulent viscosity hypothesis (TVH) assumes that the deviatoric Reynolds stress is proportional to the mean shear strain rate as (Karimpour, 2014):

$$-\left(\overline{u_i'u_j'}\right)+\frac{2}{3}k\sigma_{ij}=-v_t\left(\frac{\partial\overline{U_i}}{\partial x_j}+\frac{\partial\overline{U_j}}{\partial x_i}\right)=2v_t\overline{S_{ij}} \tag{4.8}$$

where $k=(\frac{1}{2})u_i'^2=0.5(u'^2+v'^2+\omega'^2)$ is the turbulent kinetic energy and v_t is the turbulent eddy viscosity. The gradient-diffusion hypothesis (GDH) assumes that the turbulent density (scalar) flux is aligned with the mean density (scalar) gradient as (Karimpour, 2014):

$$-\overline{\rho'u_i'}=-k_t\left(\frac{\partial\rho}{\partial x_j}\right) \tag{4.9}$$

where k_t is a positive scalar named the turbulent eddy diffusivity. These hypotheses resolve the closure problem by decreasing the number of unknowns, but they still require a correct proposition for the turbulent viscosity and diffusivity.

Different closure schemes have been widely introduced to define the turbulent viscosity (v_t). Regarding how to solve for v_t, in terms of additional equations, these closure schemes are classified as zero-equation, one-equation or two-equation models. Zero-equation or algebraic models do not require additional partial differential equations (PDEs) for transport equations and provide a prediction of the turbulent viscosity (v_t) directly from the mean flow variables. One-equation models involve the use of one additional transport equation (usually turbulent kinetic energy) and assess the turbulent viscosity (v_t) based on the estimated turbulent quantity. Two-equation RANS closure schemes such as the standard k-ε model make use of two more transport equations for turbulence quantities to define the turbulent viscosity (v_t). However, to provide closure for the turbulent flux term in the density transport equation, most turbulence schemes make use of a turbulent Prandtl number (Pr_t) instead of defining the turbulent diffusivity (k_t) explicitly. The turbulent Prandtl number is defined as:

$$Pr_t = \frac{v_t}{k_t} \tag{4.10}$$

Due to averaging, the accuracy of RANS models is always a concern in mixing problems, and the development of more robust numerical schemes is an ongoing effort. On the other hand, LES models are now receiving more attention, especially with the advancement in computational resources. LES models have the ability to resolve the unsteady nature of large eddies and small-scale eddies. To handle this, LES uses a filtering technique to split the velocity field into mean (U) and residual (u') values. The filter size can either be determined implicitly by the numerical domain grid size or by introducing filter functions (Gullbrand and Chow, 2002). The LES equations for a 3D unsteady-state flow using Boussinesq approximation can be written as follows:

$$\frac{\partial(U_i)}{\partial t} + \frac{\partial(U_i U_j)}{\partial x_i} = \frac{1}{\rho 0}\frac{\partial(p)}{\partial x_i} + v\frac{\partial^2(U_i)}{\partial x_i \partial x_j} - g\frac{(\rho)}{\rho 0}\sigma_{i3} - \frac{\partial \tau_{ij}^{SGS}}{\partial x_i} \tag{4.11}$$

The filtered density transport equation is:

$$\frac{\partial(p)}{\partial t} + \frac{\partial(\rho U_j)}{\partial x_j} = \mathrm{k}_m \frac{\partial^2(p)}{\partial x_i \partial x_j} - \frac{\partial \chi_j^{SGS}}{\partial x_j} \tag{4.12}$$

In Equations (4.11) and (4.12), τ_{ij}^{SGS} and χ_j^{SGS} are the SGS tensor and sub-grid scale flux vector, respectively, and are defined as:

$$\tau_{ij}^{SGS} = (U_i U_j) - (U_i)(U_j) \tag{4.13}$$

$$\chi_j^{SGS} = (\rho U_j) - (\rho)(U_j) \tag{4.14}$$

With the existence of residuals, similar to RANS models, LES suffers from the closure problem and requires SGS models to resolve closure. The accuracy of LES is related to the efficiency of SGS models used to define the SGS motions. The simplest and most well-known closure method was introduced by Smagorinsky (1963). A linear turbulent viscosity (v_t) is used to model the residual motions as:

$$\tau_{ij}^{SGS} = -2v_t (S_{ij}) \tag{4.15}$$

The computational studies on the mixing of effluent (review of previous studies in this chapter) in the near-field region showed that, especially for LES models, the influence of small changes in the discharge momentum (i.e., inflow momentum) can result in large variations in the concentration field (Dejoan et al., 2007). As seen in previous studies, for effluent mixing problems, RANS models perform relatively well in replicating the geometrical and flow properties of effluent discharges and thus need more exploration in terms of the sensitivity of these models to various input parameters. Amongst researchers and professionals, RANS models are very popular for mixing applications, as they have a low CPU demand and still retain an appropriate methodology. There are some applications, such as the mixing of toxic agents, in which the models should go beyond the estimate of mean concentration and should resolve smaller scales of fluid motion. In these cases, LES models are a suitable approach, and they have been an approved method based on studies in the past few years.

4.1.2 Vertical jets

Vertical discharges have traditionally been preferred over inclined jets when there are horizontal confinements, and they are usually more suitable for deep waters where there is enough height in the water column to diffuse the concentration. This section reviews the previous CFD studies completed on vertical jets. Unlike inclined dense discharges, this section includes both positively and negatively buoyant jets. Vertical jet modeling is attractive for researchers due to its simplicity and practicality.

Fletcher et al. (1993) compared the results of three different CFD programs (CHYMES multiphase flow model, finite element program, and Harwell-FLOW3D finite volume program) for laminar fluid injection into a fixed fluid cell. The results showed that the models had good consistency. Furthermore, differently encoded features were used to examine the effects of solution options (such as difference schemes) and solution parameters (such as space and time steps).

El-Amin et al. (2010) modeled a confined turbulent buoyant discharge using the realizable k-ε model of Shih et al. (1995) and modified it to account for turbulence due to buoyancy. They also validated the numerical model with an experimental study, which included the jet temperature, axial velocities, radial velocity, dynamic pressure, centerline velocity, mass, and momentum fluxes. This study is limited because it only performed 2D modeling of the 3D experiment. Numerically, they compared the realizable k-ε model with other k-ε models (standard k-ε, RNG k-ε, and k-ω) for temperature propagation in the jet. The realizable k-ε model performed the best among all models, especially closer to the nozzle, and it satisfies some constraints in the Reynolds stresses, which makes it more suitable for this specific physics of flow. Gildeh et al. (2015) made a similar recommendation but for inclined dense discharges. They expressed an essential coefficient of the realizable k-ε model, $C\mu$, as a function of mean flow and turbulence properties, instead of assuming that it was a constant, as in the standard model. This expression (i.e., realizability) satisfies certain constraints on the Reynolds stresses, which makes this model stronger compared to the other k-ε models for mixing problems.

Abou-Elhaggag et al. (2011) also studied vertical dense discharges, with both experimental and numerical investigations. A single round outfall was modeled experimentally and numerically. Several numerical simulations were completed with the ANSYS Fluent CFD package, and the numerical results were compared to the experimental data. They considered homogeneous ambient water with no current and reported the geometrical and concentration properties of the discharge along its trajectory. A wide range of conditions was covered by studying various combinations of port diameters and concentrations of effluent salinities. Their experimental results were compared to the experimental data in the literature. However, this study only compared the numerical results with previous semi-empirical formulas, so it lacked comparison to a good experimental dataset, even to their own data. The numerical model also identified a penetration depth and the existence of multiple peaks for the brine concentration.

Yan and Mohammadian (2017) investigated laterally confined vertical buoyant jets with the OpenFOAM model. They modified the standard k-ε turbulence model with the GGDH buoyancy approach and linked the Prandtl and turbulent Prandtl numbers to the Froude number. They verified the model using experimental results (Lee and Lee, 1998) and extracted normalized concentration Gaussian profiles along the jet centerline, perpendicular to the trajectory. Chen and Rodi (1980) calculated the free jet concentration, and comparing this to the modeled confined jet concentration in this study shows that using the same Froude number gives a higher dilution at a certain point along the jet centerline in confined jets. The study had the following results:

- An assumption was made that improves the determinacy and aids in the calibration of CFD modeling: the Prandtl number P_r and turbulent Prandtl number Pt_r would be related to the densimetric Froude number Fr_d:

$$Pt_r = P_r = (0.032Fr_d + 0.89) - 1 \tag{4.16}$$

- The influence of the water surface is smaller than the influence of confinement, although the distribution of concentration along a cross-section in a confined discharge could be impacted by boundaries.
- A conclusion was made to enable engineers and researchers to quickly estimate the evolution of a laterally confined vertical buoyant discharge: the rate of discharge concentration growth is almost equal to $b_{gc}/s = 0.0938$, where b_{gc} is the concentration 1/e width (e is the Napier's constant) and that is where the jet concentration reaches 1/e of the centerline maximum concentration.

Yan et al. (2020) used the OpenFOAM model to simulate the mixing properties of vertical buoyant jets discharged from multiport diffusers. Four turbulence models were investigated: standard k-ε, RNG k-ε, k-ω, and SST k-ω. By comparing cross-sectional variations of mean axial velocities at different elevations, they observed that the RNG k-ε model performed the best of the four models. By varying the ratio of the port spacing over the port diameter, they demonstrated that the port spacing affects the dilution characteristics of the jets (i.e., the concentration decreases with increasing port spacing). This study found that when the merging point is located above the nozzle, the centerline concentration drops to approximately 20% of the initial concentration, and the port spacing effect becomes less important. Therefore, a logarithmic term was added to derive a new empirical formula, which includes the port spacing effect to describe the centerline dilution:

$$\frac{C_m}{C_0} = \alpha\left(\frac{y}{d_p}\right)^{\beta} - \gamma\ln\left(\frac{p_s}{dp}\right) \tag{4.17}$$

where d_p is the port diameter, p_s is the port spacing, and α, β, and γ are regression coefficients.

More recently Gildeh et al. (2022) studied on the worst-case scenario of the dilution and mixing of vertical jets: vertical dense effluent discharges with no ambient current, in shallow water, where the jet impinges the water surface. This scenario provides conservative design criteria for such outfall systems. The numerical modeling of such jets has not been investigated before and this study provides novel insights into simulations of vertical dense effluent discharges in shallow waters. Turbulent vertical discharges with Froude numbers ranging from 9 to 24 were simulated using OpenFOAM. An RSM was applied to characterize the geometrical (i.e., maximum discharge rise Z_m and lateral spread R_{sp}) and dilution μ_{min} properties of such jets. Three flow regimes were reproduced numerically, based on the experimental data: deep, intermediate, and impinging flow regimes.

4.1.3 Horizontal jets

There are two categories of horizontal jets: offset jets and wall jets. Wall jets are attached to a solid boundary (usually at the bottom), and offset jets are elevated away from the bottom and experience no effects from the wall. This section covers numerical studies on both types to date.

Huai et al. (2010) simulated horizontal buoyant wall jets in three dimensions by applying a realizable k-ε model to analyze results for temperature dilution, centerline trajectory, and cling length at various sections, as well as jet velocity. They compared their results to those of Sharp (1975) and Verhoff (1963). A linear relationship was found for the cling length as a function of the densimetric Froude number: $L/D = 3.2Fr_d$. An exponential relationship was found for the jet centerline velocity.

The conclusions from Huai et al. (2010) included the following. (i) Temperature dilution is related to the nozzle diameter D, the distance of the jet trajectory to the nozzle x, and the densimetric Froude number Fr_a. The temperature dilution in the wall jet region decays according to Equation (4.18). (ii) The central surface velocity profiles resemble that of a turbulent wall jet and generally agree with the classical curve of wall jets. (iii) The velocity profile shows strong similarity in the vertical plane beyond an $x > 5D$ distance from the nozzle, fitting to a Gaussian shape. (iv) The centerline velocity decay is related to the densimetric Froude number and Fr_d and the nozzle diameter D, and fits Equation (4.19), where U_0 is the velocity at the source and U_{m0} is the centerline maximum velocity.

$$S = 0.0725x / D + 0.85 \tag{4.18}$$

$$U_0 / U_{m0} = 0.65x / \left(DFr_d^{0.5}\right) + 0.3 \tag{4.19}$$

Li et al. (2011) used the LES model to study, for the first time, the interaction between a parallel offset jet and a plane wall jet; the large eddies were simulated directly, and the small eddies were obtained with the Dynamic Kinetic energy Subgrid-scale Model (DKSM) and the Dynamic Smagorinsky-Lily Model (DSLM). In predicting the turbulent intensity and the mean stream-wise velocity, the agreement was good between the numerical results and the experimental data, especially for DKSM. They also analyzed some aspects of the turbulence mechanism, such as the function of correlation, the velocity Probability Density Function (PDF), and the coherent structures. In some locations, the mean stream-wise velocity profiles were similar a certain distance after merging of two jets. Near the jet exit, the velocities along the centerline of the two jets were negative. The stream-wise velocities and turbulent intensity both became positive and rapidly increased to the maximum after a stagnation point, and then they gradually decreased downstream; this shows that the interaction of jets is dominant near the exit. Finally, tracer characteristics were used to demonstrate the dilution by

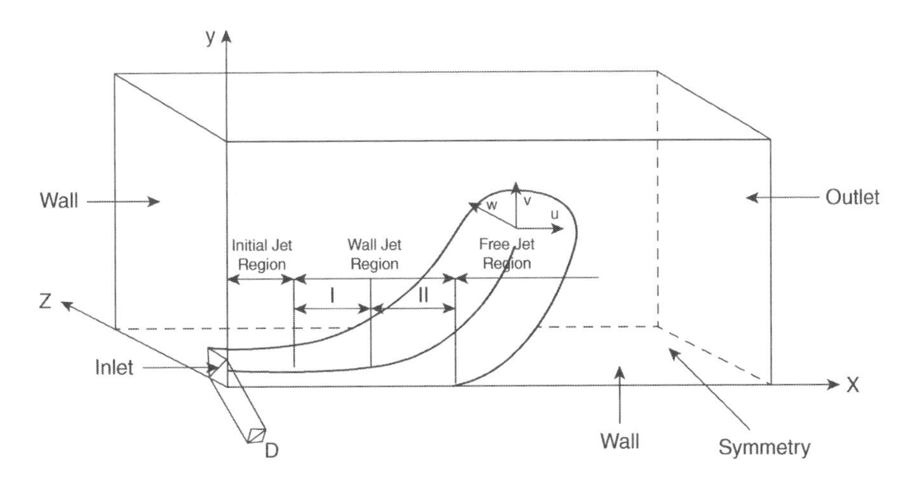

Figure 4.4 Schematic view of the model by Gildeh et al. (2014).

the dual jet. The constant concentration was maintained in both the stream-wise and wall normal directions close to the exit because of the interaction between the two jets and the presence of the wall. After the two jets merge completely, the concentrations (C/C_0) are distributed in the wall normal direction parabolically, while the maximum value decreases along the stream-wise direction linearly. The concentration (C/C_m) profiles were similar in the region with a steady C_0 value.

Horizontal wall jets (Figure 4.4) were examined by (Gildeh et al., 2014), using a finite-volume model (FVM) and several turbulence models. Their study aimed to discover whether the RANS models were suitable for predicting the velocity and concentration properties of wall jets. They also aimed to identify the best performing turbulence model. The jets were modeled with OpenFOAM, and the resulting turbulence models were compared to those from the numerical studies of Huai et al. (2010) and Sharp (1975) ($Fr_d = 11.61–42.33$). The linear (standard k-ε, RNG k-ε, realizable k-ε, and SST k-ω) and Reynolds stress (Launder-Gibson and LRR) turbulence models were tested in their study. This was the first study for wall jet modeling that compared several turbulence models.

As the fluid leaves the nozzle of a wall jet attached to the horizontal wall and discharges effluent into a water body, water entrains the jet from all directions except for the wall region. Therefore, there is a higher pressure exerted on the top of the jet than on the wall, and the jet stays on the wall up to the point where the suction pressure from above decreases, and the buoyancy force exceeds the pressure difference. There are four regions for wall buoyant jets: (i) Initial Jet Region: from the inlet to the point where the velocity profile is equal to the maximum initial velocity, and nearly uniform; (ii) Wall Jet Region I: from the end of the Initial Jet Region to the point where the jet centerline departs from the horizontal and starts rising; (iii) Wall Jet Region II: from the end of Wall Jet Region I to the point where the outer layer of the jet rises up off the floor; and (iv) Free Jet Region: starts after the Wall Jet Region. These regions are shown in Figure 4.4.

The cling length for thermal wall jets is the distance from the nozzle to the location where the floor temperature reaches $(T - T_a)/(T_o - T_a) = 3\%$ (Huai et al., 2010).

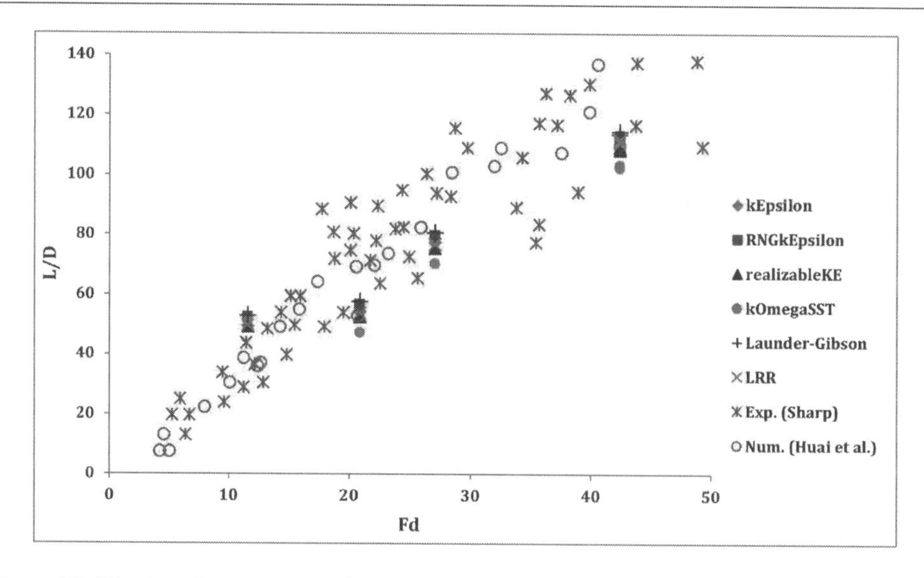

Figure 4.5 Cling length comparison of numerical vs. experimental results (Gildeh et al., 2014).

Figure 4.5 shows the numerical cling length results from Gildeh et al. (2014) compared to the experimental data and numerical results of Huai et al. (2010).

The results from Gildeh et al. (2014) agreed other researchers' numerical data and the experimental data. For larger Froude numbers, the cling length values reported by Sharp (1975) were smaller than those from Huai et al. (2010), though they align well with the numerical results obtained by Gildeh et al. (2014). Table 4.3 shows the relationships between L/D and F_d for each turbulence model evaluated in Gildeh et al. (2014).

All of the turbulence models estimated a smaller L/D than the experimental data. RNG k-ε had the closest cling length value to the experimental result, and SST k-ω had the smallest.

Predicting jet trajectories is a key factor in outfall design to predict the path that a jet travels from the exit point until it reaches the surface. This is particularly critical in locations where the receiving water is not deep enough to dilute the effluent completely. Figure 4.6 shows trajectories from several studies, including that of Gildeh et al. (2014). It seems that the trajectory results from the family of k-ε turbulence models are much more accurate than the SST k-ω model.

Gildeh et al. (2014) obtained detailed results on stream-wise velocity profiles (Figure 4.7). The stream-wise (x-y) velocity profiles of the buoyant wall jet centerline were extracted from various simulations. The velocity field results were obtained for

Table 4.3 Cling length relationship for the various turbulence models of Gildeh et al. (2014)

Turbulence model	*Standard k-ε*	*RNG k-ε*	*Realizable k-ε*	*SST k-ω*	*Launder-Gibson*	*LRR*	*Experiment (Sharp and Vyas, 1977)*
Cling length	$L/D = 2.68Fr_d$	$L/D = 2.76Fr_d$	$L/D = 2.65Fr_d$	$L/D = 2.50Fr_d$	$L/D = 2.79Fr_d$	$L/D = 2.70Fr_d$	$L/D = 3.2Fr_d$

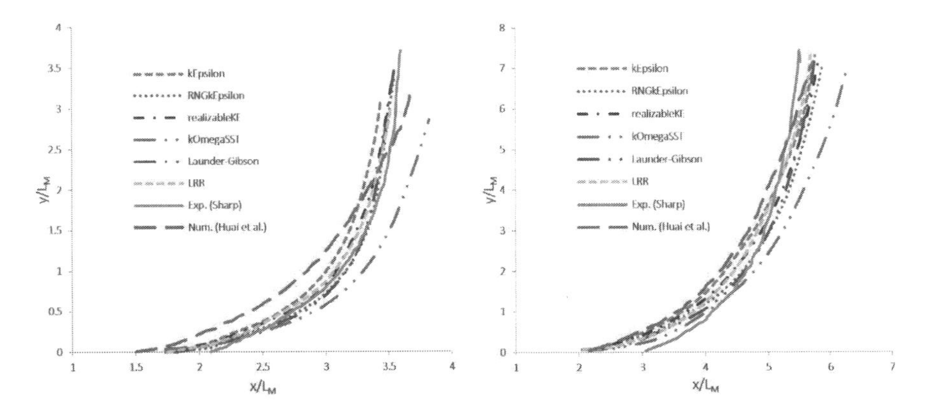

Figure 4.6 Centerline trajectory. (left) Fr_d: approximately 12 and (right) Fr_d: approximately 20 (Gildeh et al., 2014).

different jet cross-sections along the x-direction (various values of x/D) at the symmetry plane. In Figure 4.7, the variables are as follows: U_m is the x-direction velocity (along y at the central plane), U_{m0} is the maximum of U_m, with the ordinate of y, and $y_{m/2}$ is the velocity-half-height, which is the height where $U_m = U_{m0/2}$. All profiles for stream-wise velocity are self-similar and in good agreement with the data by Law and Herlina (2002), as plotted in Figure 4.7. The equation developed by Verhoff (1963), which is suitable for 2D wall jets, is also plotted along with other data for comparison. Verhoff's equation reads:

$$\frac{U_m}{U_{m0}} = 1.48\left(\frac{y}{y_{m/2}}\right)^{1/7}\left[1 - erf\left(0.68\frac{y}{y_{m/2}}\right)\right] \tag{4.20}$$

Profiles of various values of x/D showed good similarity, as seen in Figure 4.7. Buoyancy-induced instabilities justify the larger deviations in the farther x/D locations as they mostly happen at higher elevations (i.e., $y/y_{m/2} > 1$) where the momentum forces dissipate, and buoyancy forces get stronger. The literature often reports the velocity self-similarity profiles at the central plane for both experimental and numerical studies, although usually without presenting results for the offset measurement from the centerline.

Experimental data for offset velocity profiles were first reported by Law and Herlina (2002) for two offset sections, $z/D = 1.818$ and $z/D = 3.636$. Gildeh et al. (2014) compared numerical results with Law and Herlina (2002) as well as Verhoff's equation, as shown in Figure 4.8, in which $y_{m/2}$ is the local length scale and U_{ms} is the maximum velocity for the offset sections. As seen in the figure, the numerical results (current study) for $z/D = 3.636$ do not agree well with Verhoff's curve in the area close to the nozzle ($x/D = 5$ and $x/D = 10$). This is primarily due to the development of the jet in the tank width (Figure 4.9). As z/D increases, the jet may not yet develop at the values along the width of the tank, and thus the scatters show less self-similarity.

Mohammadian et al. (2020) investigated the temperature results from Gildeh et al. (2014) further. The dilution contours for temperature at the plane of symmetry

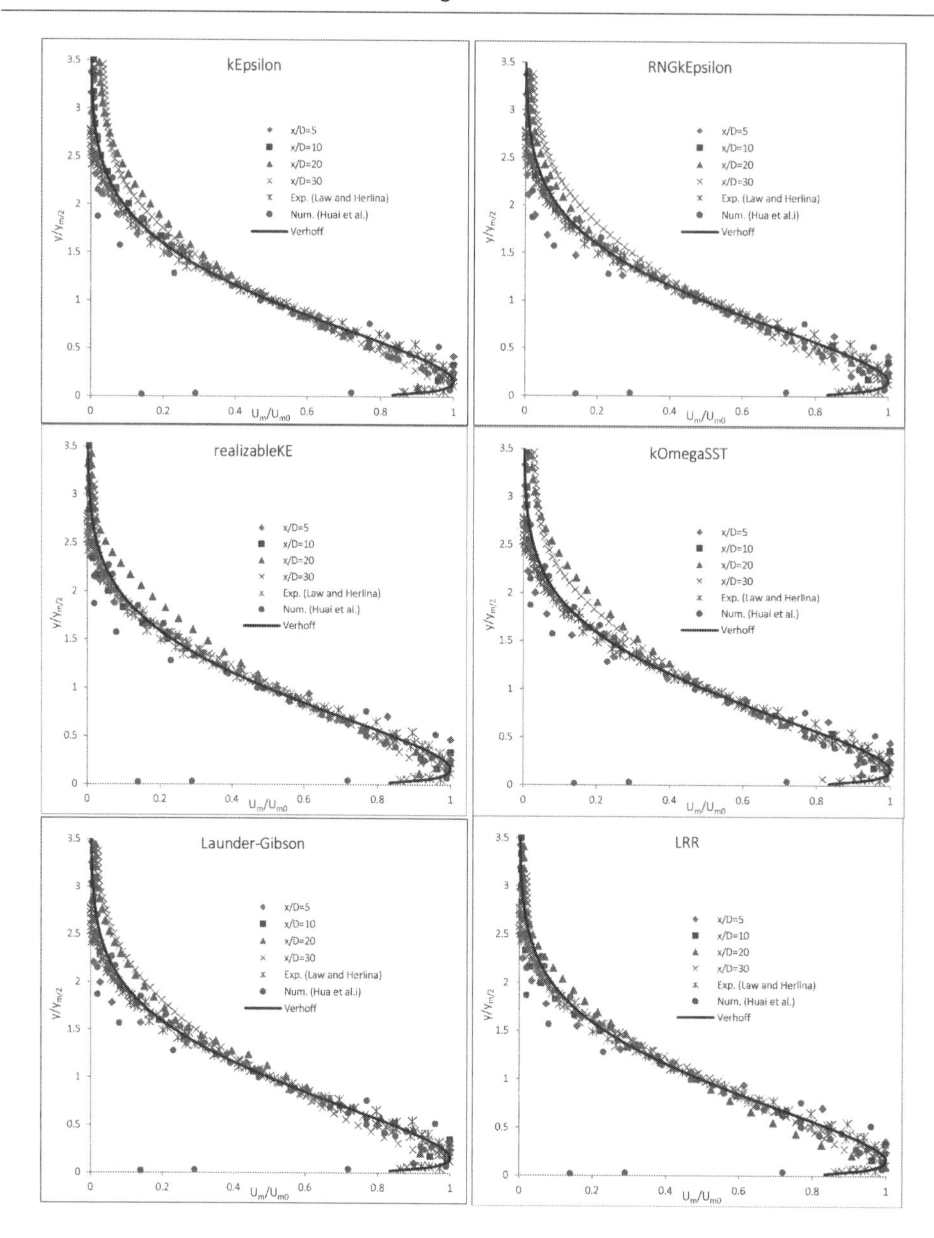

Figure 4.7 **Self-similarity of stream-wise velocity profiles for various turbulence models (Gildeh et al., 2014).**

($z = 0$) are plotted for realizable k-ε in Figure 4.10 for dilution rates of 12, 15, 20, 30, and 60. The innermost contour line reads a dilution of $S = 12$, and the outermost contour line corresponds to a dilution of $S = 60$. Dilution would obviously increase with distance from the nozzle, and it depends on both discharge and receiving water properties (such as discharge diameter D, densimetric Froude number Fr_d, and ambient water depth H_a, etc.). It was observed that the effect of distance (the path that the jet

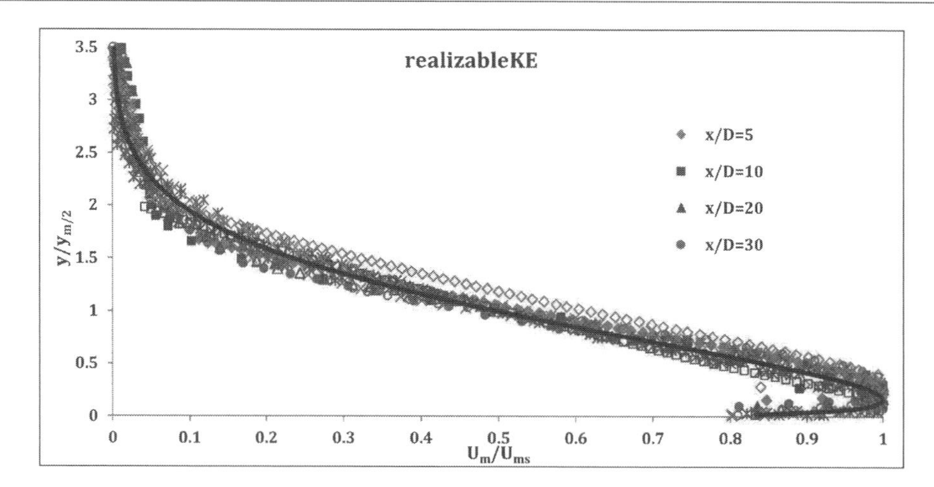

Figure 4.8 Velocity at offset sections $z/D = 1.818$ and 3.636. Solid fill scatters are for $z/D = 1.818$ and the no-fill scatters are for $z/D = 3.636$ at the x/D values on the plot.

travels through) on the dilution factor was larger than the effect of the Froude number at the nozzle.

Temperature profiles (self-similar profiles) would behave the same (i.e., Gaussian shape) for various Froude numbers. Figure 4.11 presents the Gaussian temperature distribution at different cross-sections for three different cases (i.e., three different Froude numbers), while bulked together. As shown, the temperature profiles, similar to the velocity profiles, seem to be independent of the Froude number.

Chan et al. (2014) established a particle tracking model for horizontal sediment jets through numerical simulation to predict particle settling and deposition in sand-laden jets. Consistent with basic experiments using mesh-generated turbulence and CFD calculations, the model predicted that the apparent settling velocity could be reduced to 30% of the hydrostatic settling velocity. The model prediction results were in good agreement with the experimental data of the seabed sedimentary profile. For the first time, the cross-sectional sediment profile of a horizontal sediment transport jet has been satisfactorily predicted.

Ghaisas et al. (2015) used a sigma SGS eddy viscosity model to perform a LES simulation of turbulent horizontal buoyant and nonbuoyant jets with several Reynolds (*Re*)

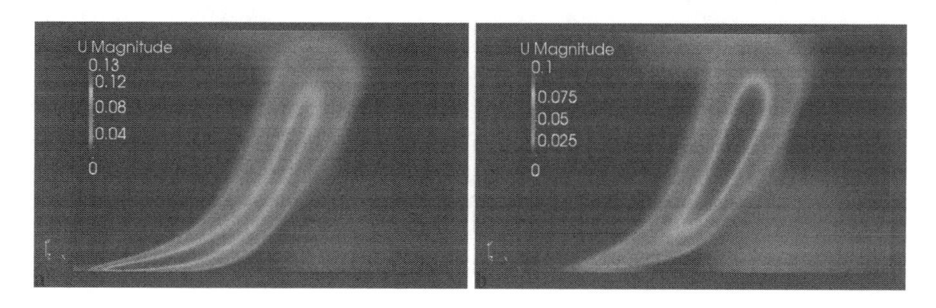

Figure 4.9 Velocity contour plots at two offset sections. (a) $z/D = 1.818$ and (b) $z/D = 3.636$.

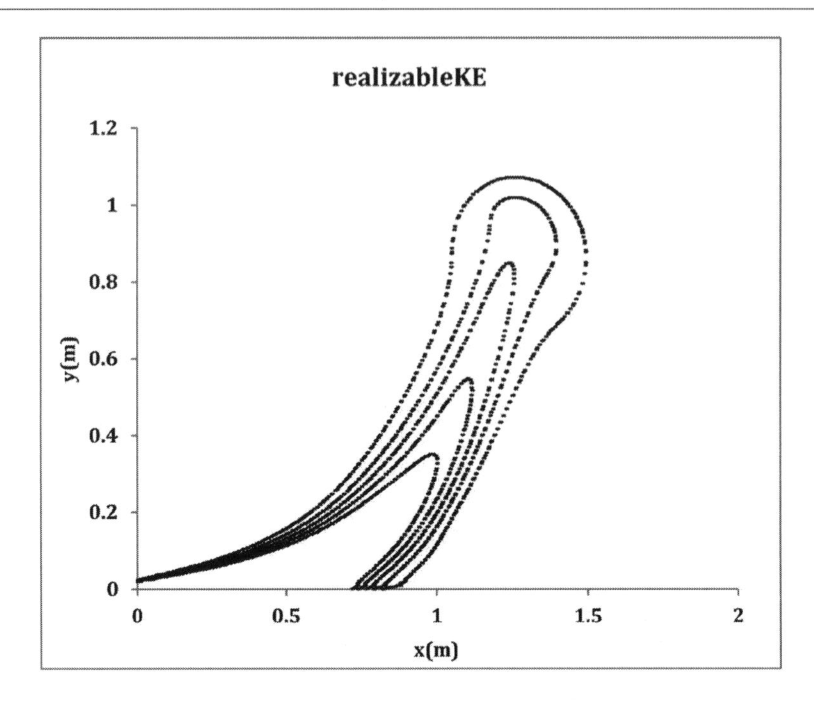

Figure 4.10 Temperature dilution contours at the plane of symmetry. Dilution rates are 12, 15, 20, 30, and 60 (realizable *k*-ε turbulence model).

and Richardson (*Ri*) numbers in stratified ambient water. Then, the effects of varying *Rt* (to characterize density differences) and *Re* (to characterize injection momentum) was studied. Turbulent production and dissipation rates were found to be asymmetric in the mid-vertical plane but symmetric in the horizontal plane. The results of Ghaisas et al. (2015) on jet centerline velocity decay, mean velocity self-similarity, radial spread, and turbulent fluctuations were in good agreement with previous experimental results.

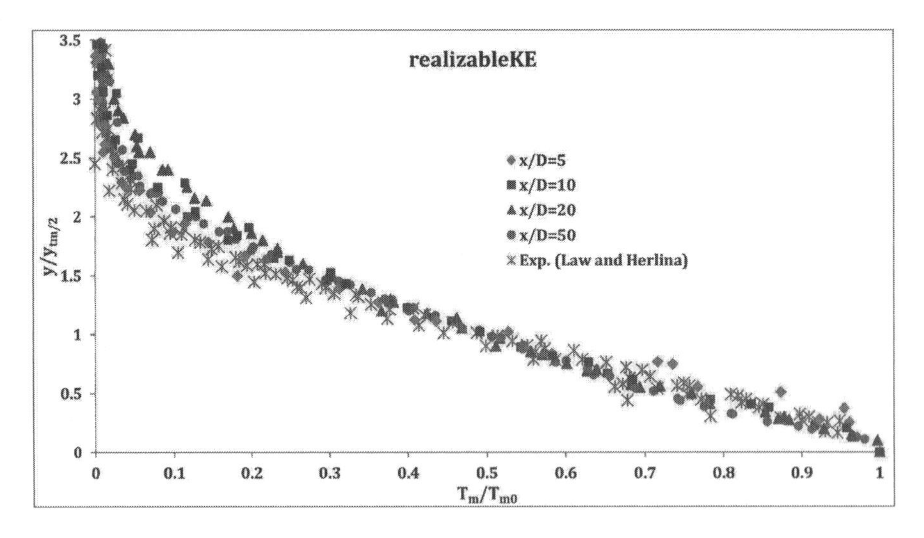

Figure 4.11 Stream-wise self-similarity temperature profiles for three cases at various cross-sections.

Using the results from their LES model and examining the instantaneous velocities in the jet studies, they realized that the jet close to the nozzle would behave the same as the developed jet farther from the discharge point in the stratified ambient water. They also identified the stable and unstable stratification regions and how the turbulent vortex rings and *Rt* are related to them.

Circular, square, and rectangular nozzles were considered in a CFD study by Mohammadaliha et al. (2016) covering the effects of nozzle geometry on turbulent offset jet development in the near-field region. Turbulence models that were studied include the standard k-ε model (Launder and Spalding, 1974), the realizable k-ε model (Shih et al., 1995), the Launder-Sharma k-ε model (Launder and Sharma, 1974), and the Yang-Shih k-ε model (Yang and Shih, 1993). The Yang-Shih k-ε turbulence model came out on top for predicting the jet properties in a comparison of these turbulence models (with a *Re* number of approximately 8,500) with previous experimental studies. Square-shaped offset jets seem to spread more in the wall normal and lateral directions, and result in more efficient mixing with the surrounding fluid than circular offset jets. However, the maximum shear stress on the adjacent wall in the case of the square-shaped nozzle was slightly higher than that in the circular nozzle case.

Zhang et al. (2015) studied turbulent circular wall jets both experimentally and by the LES model; then, they compared those results with the numerical results from two RANS models: the standard k-ε and standard k-ε models, with enhanced wall functions. Then, Zhang et al. (2015) tested the grid convergence using the Grid Convergence Index (GCI), with low-resolution (0.9 million cells) followed by high-resolution (3.1 million cells) and ending with both the velocity and concentration properties of the jets. The results of Zhang et al. (2015) showed that LES is better than both RANS models for reproducing the scalar mixing and kinematic characteristics. Their LES results generally had better agreement with the experimental data, although they did underpredict the span-wise velocity profiles away from the jet centerline. Vorticity distribution and turbulence intensities (u'/U, with u' being the root-mean-square of turbulent velocity fluctuations and U being the mean velocity) were also extracted and compared to the past experimental data, which showed better agreement of the LES model rather than RANS models. In the region far from the centerline, the span-wise turbulence intensity was observed to decrease faster, which was likely due to the inadequacy of the Smagorinsky SGS model for low-turbulence intensity situations in those regions. In the plane of symmetry, the y-direction (lateral – perpendicular to the plane of symmetry) vorticity, caused by the presence of the wall, dominated the vorticity distribution. In the horizontal plane, the primary contributor was observed to be the z-direction (vertical – perpendicular to the horizontal plane) vorticity due to the jet-flow shear layer in the ambient water.

In a recent study, Alfaifi et al. (2019) studied offset buoyant jets with various properties of the discharges and ambient water, experimentally and numerically. Discharges were set to be both thermal and nonthermal, but always positively buoyant. The ambient water was stagnant, and they used a PIV system to collect the experimental data. All comparative experiments were conducted with the same densimetric Froude numbers (Fr_d, ranging from 9.9 to 29.8) and density differences ($\Delta\rho$, ranging from 5.1 to 17.41). Three RANS turbulence models were adopted for their numerical study: standard k-ε, realizable k-ε, and buoyancy-modified k-ε. They concluded that the realizable k-ε model was more successful in predicting the discharge trajectories. The main finding of this study was that while using different combinations of parameters in discharge (salinity versus temperature) to keep the properties of the jet the same (the same values of Fr_d and same $\Delta\rho$), the trajectory and mixing characteristics of the jets would be different in the same ambient water. Therefore, it is important to not only examine the relative buoyancy between discharge and ambient water, but also the properties of discharge such as salinity and temperature, which could be very important in the overall mixing efficiency of the jets.

4.1.4 Surface discharges

Surface discharges of effluent into water bodies are less common due to lower mixing efficiency with the ambient water (i.e., the top portion of the jet/plume does not experience ambient water entrainment, especially close to the nozzle where the momentum effect is higher). This could be the reason that surface discharges have been less studied both experimentally and numerically. There is no CFD modeling of surface discharges to the best knowledge of the authors. The CORMIX3 empirical-based model was developed for surface discharge outfalls, which is not considered a CFD tool. This gap in the literature could be bridged with the following suggestions for both experimental and numerical modeling of surface discharges:

- Surface discharges using different channel geometries in calm ambient water.
- Surface discharges using different channel geometries in co-flow ambient water.
- Surface discharges using different channel geometries in cross-flow ambient water.

4.2 Future steps in modeling of effluent discharges

Although the mixing characteristics of effluent discharges have been significantly studied in the past several decades, the simulation of effluent discharges using modern mathematical and computational techniques is still in its infancy. Future studies are required to further improve the practice of modeling effluent discharges. This section provides some suggestions for future steps in modeling of effluent discharges, including improving turbulence modeling approaches, 3D simulations for effluent discharges in stratified environments, modeling the Coriolis effects on effluent discharges, simulating reaction processes, simulating effluent discharges in waves, considering the interactions between discharges or between fluids and solids, and coupling with modern machine learning algorithms.

4.2.1 Turbulence modeling

The recent advances in numerical and computing techniques provide a new approach for simulating wastewater jets as a three-dimensional phenomenon, and the three-dimensional numerical modeling method has become popular in the past several years. Although the numerical modeling technique has become relatively mature due to the continuous development of computing technology and resources, the issues regarding turbulence modeling for turbulent jets or plumes has not been completely solved, and further research is still required, which can be summarized in the following three aspects: selection of the most suitable turbulence model, improvement of the turbulence models, and surrogate models for turbulence modeling.

Different turbulence models have different performance in different applications regarding water effluents. Kheirkhah Gildeh et al. (2014) simulated turbulent buoyant wall jets in stationary ambient water, and the results showed that the realizable k-ε model performed the best among the tested turbulence models. Yan and Mohammadian (2019) simulated and tested the performance of the standard and re-normalization group (RNG) k-ε model on multiple inclined dense jets using a fully 3D numerical model, and the results showed that the RNG k-ε model performed better than the standard k-ε model. In addition, Yan et al. (2020a) used the (RNG) k-ε model to simulate a plane inclined jet in a linearly stratified environment, and the results proved the model's ability in modeling inclined jets in a linear stratified environment. Yan et al. (2020b) studied the mixing and dilution characteristics of vertical buoyant jets for four different turbulence models, including the standard k-ε, renormalization group (RNG) k-ε, standard k-ω, and k-ω shear stress transport (SST) models. The results showed that the predictions of the RNG k-ε model were the most accurate. In these studies, the optimal models used for different jet performance studies were not the same, so it is necessary to study other scenarios to summarize the optimal turbulence model in different situations.

The existing turbulence models are not necessarily the optimal models, so there is still room for improvement of turbulence models for different problems with respect to jets and plumes. Yan and Mohammadian (2017) reported an improvement for the widely used k-ε model. The standard k-ε turbulence model was originally developed for uniform density flow, but turbulent buoyant jets are driven by buoyancy due to density changes. Therefore, the k-ε model must be modified to account for the effect of buoyancy on turbulence. The k-ε GGDH turbulence model was developed by incorporating the buoyancy source term based on the generalized gradient diffusion assumption (GGDH) into the k-ε turbulence model. By comparing the results of the improved k-ε GGDH turbulence model for laterally confined buoyant jets with previous experimental results, the results showed that the buoyancy-corrected k-ε model performs better in simulating the mixing characteristics of laterally confined buoyant jets. This study demonstrated that improving the turbulence model is very promising for enhancing the practice of effluent modeling.

Currently, data-driven machine learning methods have been widely applied to study the fluid dynamics of complex flows. For example, Ajeel Fenjan et al. (2016) predicted the flow field of a curved open-channel flow using a multilayer perceptron (MLP). Kochkov et al. (2021) developed a data-driven model to greatly enhance the efficiency of computations. Surrogate models were obtained by training the deep learning model, which can effectively improve the efficiency of turbulence modeling. Duraisamy et al. (2015) studied the potential of using machine learning approaches to improve the accuracy of closure models for turbulent and transitional flows. Tracey et al. (2015) demonstrated the potential of machine-learning algorithms in enhancing and/or replacing the traditional turbulence models, such as Spalart-Allmaras. Using the gene-expression programming algorithm, Zhao et al. (2020) developed an explicit Reynolds-stress model directly implemented into RANS equations. Novati et al. (2021) used multiagent reinforcement learning (MARL) to obtain more accurate turbulence models. Mohan and Gaitonde (2018) employed proper orthogonal decomposition (POD) and long short-term memory (LSTM) architecture to develop a reduced-order model for turbulent flow control. Lui and Wolf (2019) developed a flow field predictive method using a deep feed-forward neural network (DNN). The traditional turbulence models incur heavy computational costs, but machine learning based surrogate models can replace the original turbulence models and improve the model efficiency. However, using such techniques to improve the practice of turbulence modeling of jets or plumes requires further investigation.

4.2.2 Effluents in stratified environments

At present, most studies on wastewater discharge focus on the discharge in a uniform environment, in which the received water has a uniform density. However, the receiving fluids in many practical simulations are stratified, and the stratification may significantly affect the mixing processes of the wastewater jets. For example, the continuous rise of buoyant jets may be prevented by stratification, forcing jet horizontal diffusion. Therefore, the jet theory in homogeneous environments is not suitable for jets in stratified environments. The effluents in stratified environments are less understood, so it is very important to further study the topic to obtain a better understanding.

Currently, the most popular methods for studying wastewater jets in stratified fluids are laboratory and physical experimental methods. These methods are the basis for studying this phenomenon, but generally these methods are expensive and time-consuming. Therefore, the development of mathematical tools as a supplementary method is beneficial. Simple analytical or theoretical methods have been widely used in wastewater jets, but their application in stratified environments is limited due to their rather complex mechanisms.

Another promising approach to study jets in stratified fluids is numerical simulation, and the application of numerical simulation in wastewater jets has been widely reported in recent years. Because jets in stratified environments are more complex than jets in uniform environments, the numerical simulation of jets in stratified environments is only in its initial stages. Simulations of a vertical buoyant wall jet discharge into a linear stratified environment was reported by Zhang et al. (2018). Two merging turbulent forcing plumes in a linear stratified fluid was simulated by Lou et al. (2019). Both studies were simulated using commercial software FLUENT. In both studies, the jets were injected vertically into the receiving body of water; however, in many practical applications, the jets are discharged at an angle. At the same time, since the hydraulic properties of inclined jets are different than vertical jets, the ability of mathematical models to simulate inclined jets in stratified fluids must be evaluated. Yan et al. (2020c) evaluated a numerical model to simulate inclined jets in a linearly stratified environment. However, the studies were based on two-dimensional simulations for relatively simple cases, so simulations for more complex cases should be conducted in future studies.

4.2.3 Effluents in rotating fluids

Savage and Stuart (1975) studied the effects of rotation on the jet path, which provided a theoretical framework for modeling jet path with the Coriolis effects considered, and the path of a turbulent jet originating from a circular orifice in deep water was predicted theoretically, which matched with the experimental results very well.

However, in previous studies, only a few considered the effects of rotation on jet flow, especially in numerical studies on effluents. To simplify the numerical model and reduce the computational time, the Coriolis term is usually ignored. However, for large water bodies, the Earth's rotation is an important factor that affects the characteristics and flow structures of jets and plumes. Therefore, it is necessary to deepen the research on the influence of rotation on jets in the future, and add the influence of rotation parameters into numerical models to improve the simulation accuracy.

4.2.4 Reaction processes

In the process of mixing, transport, and reaction of pollutants in rivers, concentration dilution occurs through physical, chemical, and biological processes. The dilution rate reflects the speed of degradation of pollutants in water, which is an important parameter for studies of water quality. It has played an important role in the formulation of the total emission control plan. At the same time, in the work of water resources protection, the accuracy and rationality of prediction of pollutant concentration, the design of pollutant control schemes and water quality management plans are significantly affected by these parameters. However, these parameters are significantly affected by all of the processes. The existing simulations of effluents mainly focused on the advection and dispersion processes, but the reactive processes also affect the pollution dilution. Therefore, in future research, it is suggested to further strengthen the study of pollutant reactions to improve the accuracy of numerical simulations.

4.2.5 Influence of waves

Effluent discharges from industrial and natural sources often enter coastal offshore areas, and accurate assessment of their impacts on the environment is important. The jet motion is not only affected by the initial discharge conditions, but is also affected by the waves and flow dynamics in seawater. It is therefore important to study the flow characteristics of the jets and plumes subjected to waves. Many jets discharged in marine systems have been conducted, but only a few studies dealt with jet-wave interactions. Although the stagnant environmental conditions are interesting, the existence of waves or currents is very common in coastal water bodies, and stagnant conditions almost never appear in real coastal waters. Experimental studies of jets in waves have been conducted but numerical modeling of jets and plumes in waves has rarely been reported, which requires future studies.

4.2.6 Influence of interactions

When a buoyant jet is discharged into the receiving water and there is a sufficient distance between the jet and any other jet or structures, the jet is known as a free jet. In fact, in real cases, the jets are often affected by nearby structures or jets, such as a jet from a submarine outfall pipeline that is laid in a dredged trench (Lee and Lee, 1998). Yan et al. (2017) carried out a numerical study on laterally constrained vertical buoyant jets, and also studied the effect of port spacing on mixing characteristics of multiple vertical buoyant jets (Yan et al., 2019). It can be seen that jet-jet or jet-solid interactions have an evident impact on the flow structures and physical characteristics of the jets. Thus, in the future, research on jets under different interactions needs to be conducted.

4.2.7 Machine learning approaches

Artificial intelligence (AI) is a new way of studying engineering problems, and its application in water resources research has been widely reported in recent years. Automatic model development using training data without predefining the model structure is the most important advantage of artificial intelligence technology. First, the hidden relationship between the detection variables can be detected by it; second, the error caused by some predetermined assumptions can be eliminated by it; third, it is fast and efficient. With more training data available, AI models can be continuously improved or extended quickly. With the rapid development of data collection technology, the use of AI techniques to develop and improve models is becoming a mainstream trend. However, the application of AI, or more specifically, machine learning algorithms, in predicting the mixing and transport properties of effluent is still in its infancy, which requires further studies.

4.3 Conclusions

Over the past two decades, due to the improvement of computational resources and computing techniques, research and applications of CFD modeling of effluent mixing and diffusion has significantly increased. Although effluent discharges have already been extensively studied, the existing studies have primarily focused on relatively simple cases. The latest advances in experimental devices such as PIV and LIF systems, sampling techniques in field research and the availability of open source CFD tools have opened the door to more realistic CFD modeling of effluent discharges. On the basis of an extensive literature review on CFD modeling of effluent discharges, the following conclusions can be drawn:

- Most of the existing studies on numerical modeling of effluent discharges focused on laboratory scale studies. To validate models, the trajectory, dilution, and velocity characteristics were studied in detail and compared to experimental data.
- The RANS and LES turbulence models are popular in existing studies on numerical modeling of effluent discharges.
- Vertical jets are also very popular in CFD research. New trends in the study of these jets include considering the environmental conditions that may affect these jets, such as lateral confinement and shallow water conditions, where the jet is attached to the top boundary. Vertical jet crossflow may have a significant impact on trajectory and dilution, both of which have attracted more attention from researchers using CFD models.
- In the past few years, experimental and numerical studies have predominantly been carried out on single jets, and now more attention is being paid to the interaction of multiple horizontal jets beyond a certain distance from the discharge point.
- In cases with more complex configurations, such as multiport and rosette diffusers, there are also literature gaps. Due to the simplicity of boundary and internal conditions, previous studies mainly concentrated on the stagnant water environment. However, in order to simulate the conditions in real life more accurately, more complex environmental conditions need to be studied, such as conditions subjected to the effects of waves, wind, confluence, cross flow, and density stratification.

In the past few decades, researchers have done extensive research on the mixing characteristics of effluent discharges, but using modern mathematics and computing technology to model effluent discharges is still in its infancy. Future research needs to further improve the practice of simulated wastewater discharges. Some suggestions for modeling steps of future studies involving sewage discharge are provided below:

- The improvement of turbulence modeling methods.
- Realization of sewage discharge in stratified environment by 3D simulation.
- Simulation of the influence of Coriolis effect on sewage discharge.
- Simulation of reaction process.
- Simulation of sewage discharge in waves.
- Consideration of the interaction between discharge points or between fluid and solid.
- Coupling with modern machine learning algorithms.

References

Abou-Elhaggag, M.E., Elgamal, M., and Farouk, M.I. Experimental and numerical investigation of desalination plant outfalls in limited disposal areas. *J. Environ. Prot.* 2011, 2, 828–839.

Ajeel Fenjan, S., Bonakdari, H., Gholami, A., and Akhtari, A.A. Flow variables prediction using experimental, computational fluid dynamic and artificial neural network models in a sharp bend. *Int. J. Eng. Trans. A Basics* 2016, 29(1), 14–22.

Alfaifi, H., Mohammadian, A., Gildeh, H.K., and Gharavi, A. Experimental and numerical study of the characteristics of thermal and nonthermal offset buoyant jets discharged into stagnant water. *Desalin. Water Treat.* 2019, 141, 171–186.

Bemporad, G.A. Simulation of round buoyant jet in stratified flowing environment. *J. Hydraul. Eng.* 1994, 120, 529–543.

Bloomfield, L.J. and Kerr, R. Inclined turbulent fountains. *J. Fluid Mech.* 2002, 451, 283–294.

Chan, S.N. and Lee, J.H.W. Particle tracking modeling of sediment-laden jets. *Adv. Geosci.* 2014, 39, 107–114, https://doi.org/10.5194/adgeo-39-107-2014.

Chen, C.J. and Rodi, W. *A Review of Experimental Data of Vertical Turbulent Buoyant Jets*. Pergamon Press: Tarrytown, NY, USA, 1980.

Cipollina, A., Brucato, A., Grisafi, F., and Nicosia, S. Bench scale investigation of inclined dense jets. *J. Hydraul. Eng.* 2005, 131, 1017–1022.

Dejoan, A., Santiago, J.L., Pinelli, A., and Martilli, A. Comparison between LES and RANS computations for the study of contaminant dispersion in the MUST field experiment. *Am. Meterol. Soc.* 2007. submitted.

Duraisamy, K., Zhang, Z.J., and Singh, A.P. *New Approaches in Turbulence and Transition Modeling Using Data-driven Techniques*. AIAA Aerospace Sciences Meeting, 2015.

El-Amin, M.F., Sun, S., Heidemann, W., and Muller-Steinhagen, H. Analysis of a turbulent buoyant confined jet modeled using realizable k-ϵ model. *Heat Mass Transf.* 2010, 46, 943–960.

Fletcher, D.F., McCaughey, M., and Hall, R.W. Numerical simulation of a laminar jet flow: a comparison of three CFD models. *Comput. Phys. Commun.* 1993, 78(1–2), 113–120. https://doi.org/10.1016/0010-4655(93)90147-5.

Germano, M., Piomelli, U., Moin, P., and Cabot, W.H. A dynamic subgrid-scale eddy viscosity model. *Phys. Fluids A: Fluid Dyn.* 1991, 3, 1760–1765.

Ghaisas, N.S., Shetty, D.A., and Frankel S.H. Large eddy simulation of turbulent horizontal buoyant jets. *J. Turbulence*. 2015, 16(8), 772–808. DOI: 10.1080/14685248.2015.1008007

Gildeh, H.K. *Numerical Modeling of Thermal/Saline Discharges in Coastal Waters*. Master's Thesis, University of Ottawa, Ottawa, ON, Canada, 2013.

Gildeh, H.K., Mohammadian, A., Nistor, I., and Qiblawey, H. Numerical modeling of turbulent buoyant wall jets in stationary ambient water. *J. Hydraul. Eng.* 2014, 140(6), 04014012.

Gildeh, H.K., Mohammadian, A., Nistor, I., and Qiblawey, H. Numerical modeling of 30_ and 45_ inclined dense turbulent jets in stationary ambient. *Environ. Fluid Mech.* 2015, 15, 537–562.

Gildeh, H.K., Mohammadian, A., Nistor, I., Qiblawey, H., and Yan, X. CFD modeling and analysis of the behavior of 30° and 45° inclined dense jets—New numerical insights. *J. Appl. Water Eng. Res.* 2016, 4, 112–127.

Gildeh, H.K., Mohammadian, A., and Nistor, I. Inclined dense effluent discharge modelling in shallow waters. *Environ. Fluid Mech.* 2021, 21, 955–987. https://doi.org/10.1007/s10652-021-09805-6

Gildeh, H.K., Mohammadian, A., and Nistor, I. Vertical dense effluent discharge modelling in shallow waters. *Water* 2022, 14, 2312. https://doi.org/10.3390/w14152312.

Gullbrand, J. and Chow, F. Investigation of numerical errors, subfilter-scale models, and subgrid-scale models in turbulent channel flow simulations. In Proceedings of the Summer Program; Center for Turbulence Research; NASA Ames/Stanford University: Stanford, CA, USA, 2002, pp. 87–104.

Huai, W., Li, Z.-W., Qian, Z.-D., Zeng, Y.-H., Han, J. and Peng, W.-Q. Numerical simulation of horizontal buoyant wall jet. *J. Hydrodyn.* 2010, 22, 58–65.

Jiang, B., Law A.W.K., and Lee J.H.W. Mixing of 30° and 45° inclined dense jets in shallow coastal waters. *J. Hydraul. Eng.* 2014, 140(3), 241–253.

Jiang, M. and Law, A.W.K. Mixing of swirling inclined dense jets—A numerical study. *J. Hydro-Environ. Res.* 2018, 21, 118–130.

Karimpour, F. *Turbulence Modelling of Stably Stratified Wall-Bounded Flows*. Ph.D. Thesis, Colorado State University, Fort Collins, CO, USA, 2014.

Kikkert, G.A., Davidson, M., and Nokes, R.I. Inclined negatively buoyant discharges. *J. Hydraul. Eng.* 2007, 133, 545–554.

Kochkov, D., Smith, J.A., Alieva, A., Wang, Q., Brenner, M.P., and Hoyer, S. Machine learning accelerated computational fluid dynamics. *PNAS* 2021. doi: 10.1073/pnas.2101784118.

Launder, A. and Spalding, D.B. The numerical computation of turbulent flows. *Comput. Methods Appl. Mech. Eng.* 1974, 3(2), 269–289.

Launder, B.E. and Sharma B.I. Application of the energy-dissipation model of turbulence to the calculation of flow near a spinning disc. *Lett. Heat Mass Transfer*. 1974, 1(2), 131–137.

Law, A.W.-K. and Herlina, H. An experimental study on turbulent circular wall jets. *J. Hydraul. Eng*. 2002, 128, 161–174.

Lee, A.W.-T. and Lee, J.H.-W. E_ect of lateral confinement on initial dilution of vertical round buoyant jet. *J. Hydraul. Eng*. 1998, 124, 263–279.

Li, Z.-W., Huai, W., and Han, J. Large eddy simulation of the interaction between wall jet and O_set jet. *J. Hydrodyn.* 2011, 23, 544–553.

Lou, Y., He, Z., Jiang, H., and Han, X. Numerical simulation of two coalescing turbulent forced plumes in linearly stratified fluids. *Phys. Fluids* 2019, 31(3), 037111.

Lui, H.F.S. and Wolf, W.R. Construction of reduced-order models for fluid flows using deep feed-forward neural networks. *J. Fluid Mech.* 2019, 872, 963–994.

Mohan, A.T. and Gaitonde, D.V. A deep learning based approach to reduced order modeling for turbulent flow control using LSTM neural networks. *arXiv* 2018. DOI: 10.48550/arXiv.1804.09269.

Mohammadaliha, N. et al. Numerical investigation of nozzle geometry effect on turbulent 3-D water offset jet flows. *J. Appl. Fluid Mech.* 2016, 9, 2083–2095.

Mohammadian, A., Gildeh, H.K., and Nistor, I. CFD modeling of effluent discharges: A review of past numerical studies. *Water* 2020, 12(3), 856.

Nemlioglu, S. and Roberts, P.J. Experiments on dense jets using three-dimensional laser-induced fluorescence (3DLIF). In Proceedings of the 4th International Conference on Marine Waster Water Discharges & Coastal Environment (MWWD), Antalya, Turkey, 6–10 November 2006.

Novati, G., de Laroussilhe, H. L., and Koumoutsakos, P. Automating turbulence modelling by multi-agent reinforcement learning. *Nat. Mach. Intell.* 2021, 3(1), 87–96.

Oliver, C.J., Davidson, M.J., and Nokes, R.I. *K-ϵ* predictions of the initial mixing of desalination discharges. *Environ. Fuild Dyn*. 2008, 8, 617–625.

Oliver, C.J., Davidson, M.J., and Nokes, R.I. Behavior of dense discharges beyond the return point. *J. Hydraul. Eng.* 2013, 139, 1304–1308.

Papakonstantis, I.G., Christodoulou, G.C., and Papanicolaou, P. Inclined negatively buoyant jets 1: Geometrical characteristics. *J. Hydraul. Res.* 2011a, 49, 3–12.

Papakonstantis, I.G., Christodoulou, G.C., and Papanicolaou, P. Inclined negatively buoyant jets 2: Concentration measurements. *J. Hydraul. Res.* 2011b, 49, 13–22.

Piggott, M., Gorman, G.J., Pain, C.C., Allison, P.A., Candy, A.S., Martin, B.T. and Wells, M.R. A new computational framework for multi-scale ocean modelling based on adapting unstructured meshes. *Int. J. Numer. Methods Fluids* 2008, 56, 1003–1015.

Ramezani, M., Abessi, O. and Firoozjaee, A.R. Effect of proximity to bed on 30° and 45° inclined dense jets: A numerical study. *Environ. Process.* 2021, 8, 1141–1164. https://doi.org/10.1007/s40710-021-00533-z

Roberts, P.J.W., Ferrier, A., and Daviero, G. Mixing in inclined dense jets. *J. Hydraul. Eng.* 1997, 123, 693–699.

Savage, SB. and Sobey, RJ. Horizontal momentum jets in rotating basins. *J. Fluid Mech.* 1975, 71, 4, 755–768. https://doi.org/10.1017/S0022112075002832.

Shao, D. *Desalination Discharge in Shallow Coastal Waters*. Ph.D. Thesis, Nanyang Technological University, Singapore, 2009.

Shao, D. and Law, A.W.K. Integral modelling of horizontal buoyant jets with asymmetrical cross sections. In Proceedings of the 7th International Symposium on Environmental Hydraulics, Singapore, 7–9 January 2014.

Sharp, J.J. The use of a buoyant wall jet to improve the dilution of a submerged outfall. *Proc. Inst. Civ. Eng.* 1975, 59, 527–534.

Sharp, J. J., and Vyas, B. D. (1977). The buoyant wall jet. *Proc. Inst. Civ. Eng. Part 2 Res. Theory*, 63(3), 593–611.

Shih, T.H., Liou, W.W., Shabbir, A., Yang, Z., and Zhu, J. A new k–ϵ eddy-viscosity model for high Reynolds number turbulent flows model development and validation. *Comput. Fluids* 1995, 24, 227–238.

Smagorinsky, J. General circulation experiments with the primitive equations. *Mon. Weather Rev.* 1963, 91, 99–152.

Tracey, B.D., Duraisamy, K., and Alonso, J.J. A *Machine Learning Strategy to Assist Turbulence Model Development*. 53rd AIAA Aerospace Sciences Meeting, 2015.

Vafeiadou, P., Papakonstantis, I., and Christodoulou, G. Numerical simulation of inclined negatively buoyant jets. In Proceedings of the 9th International Conference on Environmental Science and Technology, Rhodes, Greece, 1–3 September 2005.

Verhoff, A. The Two-Dimensional Turbulent Wall Jet with and without an External Stream, Rep. No. 626, Princeton University: Princeton, NJ, USA, 1963.

Yan, X. and Mohammadian, A. Numerical Modeling of vertical buoyant jets subjected to lateral confinement. *J. Hydraul. Eng.* 2017, 143(7), 04017016.

Yan, X. and Mohammadian, A. Numerical modeling of multiple inclined dense jets discharged from moderately spaced ports. *Water* 2019, 11(10), 2077. DOI: 10.3390/w11102077.

Yan, X., Mohammadian, A., and Chen, X. Numerical modeling of inclined plane jets in a linearly stratified environment. *Alexandria Eng. J.* 2020, 59(3). doi:10.1016/j.aej.2020.05.023.

Yang, Z. and Shih, T.H. New time scale based k-ε model for near-wall turbulence. *AIAA Journal* 1993, 31(7), 1191–1198.

Zhang, S., Law, A.W.K., and Zhao, B. Large eddy simulations of turbulent circular wall jets. *Int. J. Heat Mass Transf.* 2015, 80, 72–84. https://doi.org/10.1016/j.ijheatmasstransfer.2014.08.082.

Zhang, S., Jiang, B., Law, A.W.-K., and Zhao, B. Large eddy simulations of 45° inclined dense jets. *Environ. Fluid Mech.* 2016, 16, 101–121.

Zhang, S., Law, A.W.K., and Jiang, M. Large eddy simulations of 45° and 60° inclined dense jets with bottom impact. *J. Hydro Environ. Res.* 2017, 15, 54–66.

Zhang, Z., Guo, Y., Zeng, J., Zheng, J., and Wu, X. Numerical simulation of vertical buoyant wall jet discharged into a linearly stratified environment. *J. Hydraul. Eng.* 2018, 144(7), 06018009.

Zhao, Y., Akolekar, H.D., Weatheritt, J., Michelassi, V., and Sandberg, R.D. RANS turbulence model development using CFD-driven machine learning. *J. Comput. Phys.* 2020, 411, 1–19.

Appendix

Mesh generation in OpenFOAM

A1 Mesh generation using the BlockMesh utility

A.1 Introduction

BlockMesh is a mesh generator for structured hexahedral mesh. Its control file path is "constant/polyMesh/blockMeshDict." The utility reads this user-defined file and then generates meshes according to the configurations defined in the file. A key mechanism underlying blockMesh is to decompose large and complex simulation domains into one or more three-dimensional subdomains (i.e., hexahedral blocks). Each block contains 8 vertices and 12 edges, and each vertex is located on each corner of the hexahedron. The edges of each block can be straight lines, arcs, splines, etc. Therefore, a modeler first defines the coordinates of vertices, and these vertices are then utilized to generate blocks, which in turn form the final computational domain. BlockMesh is generally used to generate grids for cases with relatively simple geometry.

A.2 Configurations

In the blockMesh steering file, blockMeshDict, the configurations can generally be split into the following sections: vertices, blocks, edges, boundary, and mergePatchPairs.

A.2.1 Vertices

This section lists all the coordinates of the vertices contained in the block, and the vertex number starts from zero.

A.2.2 Blocks

A block is defined by sequence numbers of 8 vertices, and the corresponding numbers of grids in x, y, and z directions are also defined in this section. The grid sizes can be defined using two methods: simpleGrading and edgeGrading.

- simpleGrading uses uniform scaling in all three directions. For example, the setting "simpleGrading (2 3 4)" means that the scales along the x, y, and z directions are 2, 3, and 4, respectively.
- edgeGrading gives the complete cell scale for each edge. For example, the setting "edgeGrading (2 2 2 2 3 3 3 3 4 4 4 4)" means that the scale along edges 0~3 is 2, the scale along edges 4~7 is 3, and the scale along edges 8~11 is 4.

A.3 Edges

This section defines the edges. By default, edges connecting two vertices are straight, but different types of edges can be specified in this section. Some commonly seen edge types are summarized as follows:

- arc: connecting vertices using arcs; requires a point through which the arc passes
- simpleSpline: connecting vertices using splines; requires a series of interior points
- polyLine: connecting vertices using a series of straight lines; requires a series of interior points
- polySpline: connecting vertices using a series of splines; requires a series of interior points
- line: connecting vertices using straight lines

A.4 Boundary

This section is used to define the name and type of each boundary, and the vertex numbers that form the boundary face. The vertex numbering order should follow the right-hand rule. The name of the boundaries can be defined by the users. Some common boundary types include: wall (wall), symmetry plane (symmetryPlane), periodic boundary (cyclic), inconsistent periodic boundary (cyclicAMI), two-dimensional boundary Axisymmetric boundary (wedge), and 2D boundary (empty).

A.5 mergePatchPairs

blockMesh allows users to create meshes using multiple blocks. In the case of including multiple blocks, it is necessary to deal with the connection problem between blocks. There are two ways to merge blocks:

- face matching: when connecting blocks using the face matching approach, it is not necessary to define parameters within mergePatchPairs. BlockMesh will automatically match these two patches into inner faces.
- face fusion: when using the face fusion approach, the following rules are used: the masterPatch on the main surface remains unchanged, and the coordinates of all points on it remain unchanged. If there is a gap between the masterPatch on the main surface and the slavePatch on the secondary surface, project the slavePatch on the secondary surface onto the masterPatch on the main surface to meet the surface fusion requirements; adjust the position of the nodes on the secondary surface through the minimum tolerance value, improve the node matching degree between the primary surface and the secondary surface, and remove the subtle edges smaller than the minimum tolerance. When the main and auxiliary surfaces partially overlap, the overlapping part will become an internal surface, and the nonoverlapping part will still be an external surface, and boundary conditions need to be defined. If the secondary surface slavePatch is fully integrated into the main surface, the secondary surface will be removed.

A.5.1 Tutorial 1: Vertical discharges into a T-shaped domain

This section provides a step-by-step tutorial on how to generate the mesh using the blockMesh utility for simulations of vertical discharges into a T-shaped domain. A

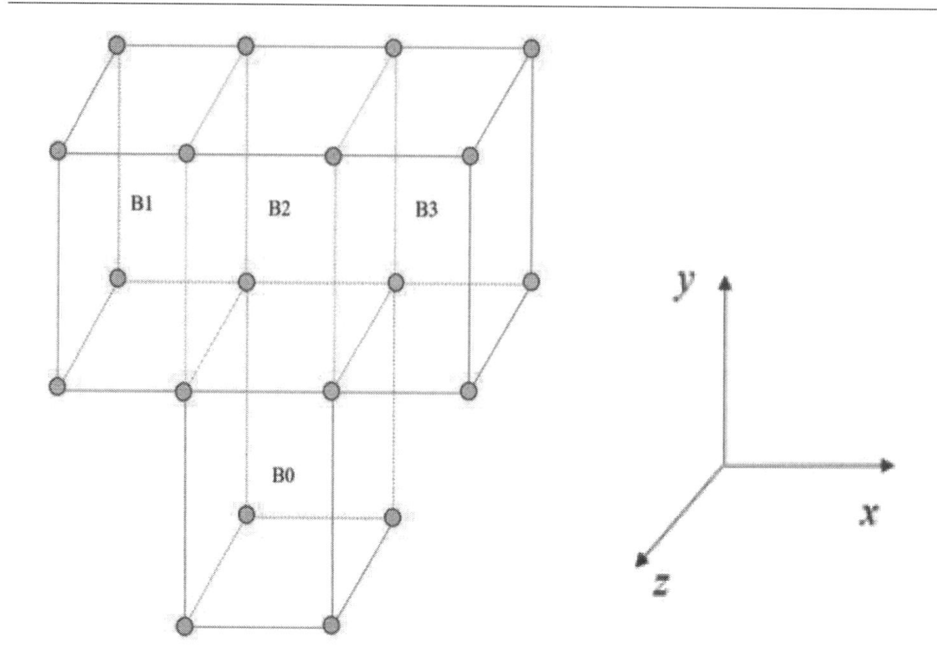

Figure A.1 Sketch of a T-shaped domain.

sketch of the problem is shown in Figure A.1. There are four blocks in the considered case, namely B0, B1, B2, and B3. In this case, the y direction is defined as the vertical direction.

It is suggested to draw a sketch beforehand using blockMesh, and the general steps for drawing a sketch are summarized as follows:

1 Define the coordinate system: define a 3D Cartesian coordinate system. In this example, the vertical axis is defined as the y-axis.
2 Draw the outline of the geometry.
3 Mark the vertices.
4 Identify the blocks.

A sample sketch is shown in Figure A.2. The file required for blockMesh to generate grids is constant/polyMesh/blockMeshDict, which primarily contains five parts: vertices, blocks, edges, patches, and mergePatchPairs.

A frequently used function in blockMeshDict is "convertToMeters." This function is used to define a scaling factor. All vertex coordinates specified in blockMeshDict will be multiplied by this factor. For example, a "convertToMeters" value of 0.001 means that all vertex coordinates will be multiplied by 0.001, and it also indicates that all coordinate values specified in blockMeshDict are in millimeter units.

The next section covers how to define the vertices. An example is illustrated in Figure A.3, and the sequence number as well as the coordinates are also shown in the same figure. As can be seen, there are 20 vertices in the example. In blockMesh, it should be noted that the numbering of vertices starts from 0, and thus the vertices are

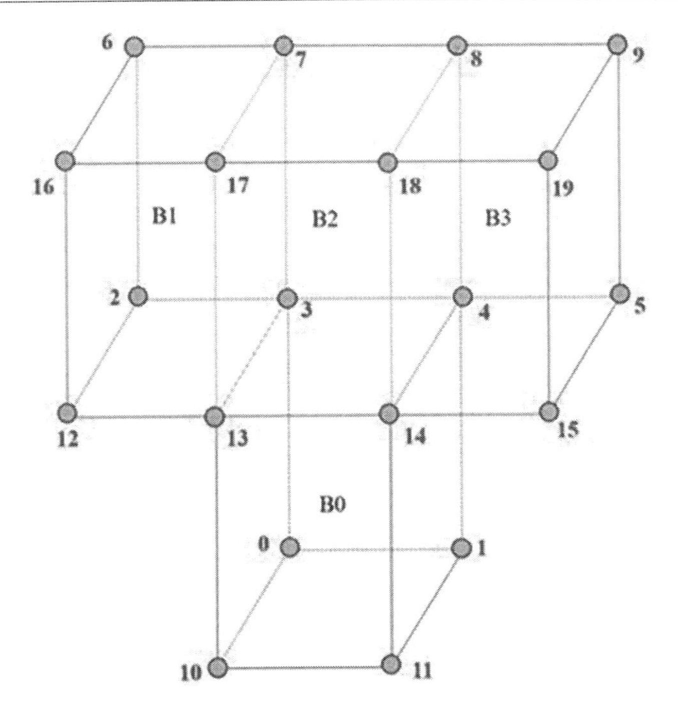

Figure A.2 Numbering of the vertices.

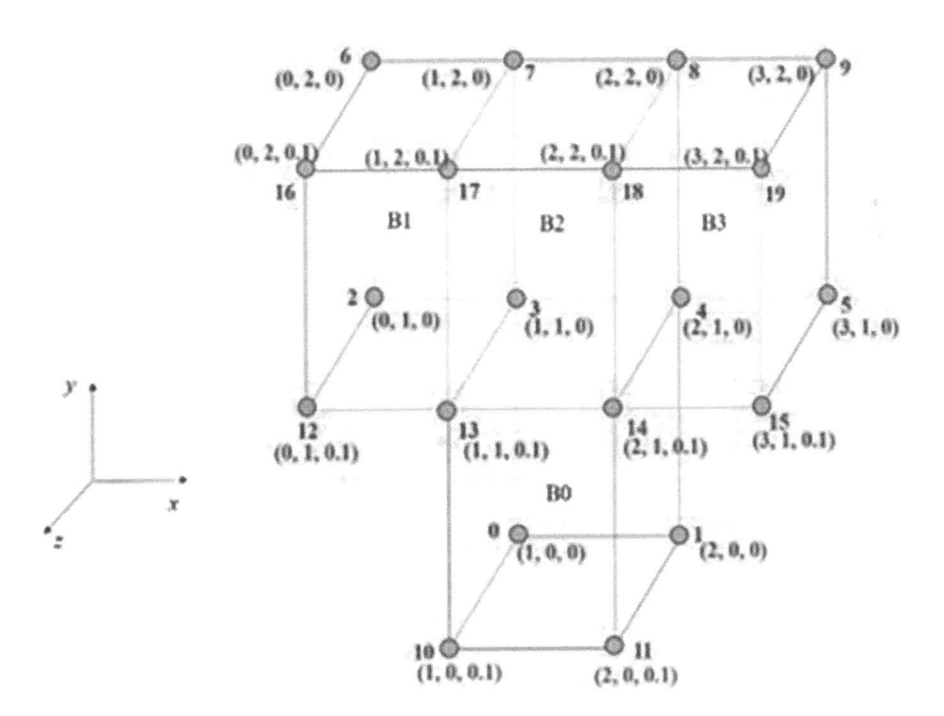

Figure A.3 The coordinates of the vertices.

numbered as 0~19. The vertices were defined by sequentially entering the coordinates for each vertex, as follows:

```
    Vertices
(
    (1 0 0) //0
    (2 0 0) //1
    (0 1 0) //2
    (1 1 0) //3
    (2 1 0) //4
    (3 1 0) //5

    (0 2 0) //6
    (1 2 0) //7
    (2 2 0) //8
    (3 2 0) //9

    (1 0 0.1) //10
    (2 0 0.1) //11
    (0 1 0.1) //12
    (1 1 0.1) //13
    (2 1 0.1) //14
    (3 1 0.1) //15

    (0 2 0.1) //16
    (1 2 0.1) //17
    (2 2 0.1) //18
    (3 2 0.1) //19

    );
```

Another important section in blockMeshDict is to define blocks. In the example shown in Figure A.3, there are four blocks, which can be named B0, B1, B2, and B3, respectively. Each block consists of eight vertices, and a block is defined by providing the sequence numbers of the eight vertices. For example, block B0 is defined by "hex (0 1 4 3 10 11 14 13)." The numbers in the second brackets specify the numbers of grids in the corresponding block. Typically, the three values indicate the numbers of grids in the x, y, and z directions, respectively. For example, "(15 15 1)" indicates 15 grids in the x and y directions, and 1 grid in the z direction. The simpleGrading command defines the ratios of grid size in the x, y, and z directions, respectively. In this example, the three values are all set as 1, indicating that the grid sizes in the block are uniform. The block section for the example illustrated in Figure A.3 is provided as follows:

```
blocks                      //Enter the blocks information
 (
    hex (0 1 4 3 10 11 14 13) (15 15 1) simpleGrading (1 1 1)
    hex (2 3 7 6 12 13 17 16) (15 15 1) simpleGrading (1 1 1)
    hex (3 4 8 7 13 14 18 17) (15 15 1) simpleGrading (1 1 1)
    hex (4 5 9 8 14 15 19 18) (15 15 1) simpleGrading (1 1 1)
 );
```

In the current example, the edges are straight lines, so the section "edges" can be kept empty, as follows:

```
edges
(
);
```

The next section defines the boundary planes. For each boundary, there are three main elements: boundary name, boundary type, and the vertices that form the boundary. In this example, the flow is discharged vertically from the bottom patch, and thus the bottom section is the inlet. The boundary is named "Inlet" in this example. It should be noted that this name is a user-defined name, and thus other names are also allowed. The type of this boundary is "patch", which is commonly used for open boundaries. The vertices are (0 1 11 10), which are the sequence numbers of the vertices that form the boundary. The present example is supposed to be a 2D case, and thus the front and back patches are defined as "empty" type. The "boundary" section in blockMeshDict is presented below.

```
Boundary
(
  Inlet
    {
          type patch;
          (
               (0 1 11 10)
          );
    }
   outlet
    {
          type patch;
          faces
          (
               (6 7 17 16)
               (7 8 18 17)
               (8 9 19 18)
          );
    }

    fixedWalls

    {
           type wall;
           (
               (0 3 13 10)
               (2 3 13 12)
               (2 6 16 12)
               (1 4 14 11)
               (4 5 15 14)
               (5 9 19 15)
           );
    }
     frontAndBack
    {
```

```
        type empty;
        faces
        (
                (0 1 4 3)
                (2 3 7 6)
                (3 4 8 7)
                (4 5 9 8)

                (10 11 14 13)
                (12 13 17 16)
                (13 14 18 17)
                (14 15 19 18)
            );
    }
);
```

The current example does not use the "mergePatchPairs", and thus the section is kept blank, as follows.

```
mergePatchPairs
(
);
```

After the blockMeshDict file is prepared, the utility can be executed via a command prompt by typing "blockMesh".

A.6 Mesh generation using the salome utility

A.6.1 Tutorial 2: A jet discharged into a channel bend

This section provides a step-by-step tutorial on how to generate the mesh using the Salome utility for simulations of a jet discharged into a channel bend. A snapshot of the final mesh is shown in Figure A.4.

It is suggested to first set the unit system. To set the unit system:

- Select the Geometry Module
- File/Properties
- Length units: m
- OK

The next step is to create the geometry, which is summarized below.

1 Straight Entrance Section (Figure A.5)

    - New Entity/Basic/2D Sketch
    - (−1,0), apply
    - (−1,12.2) apply
    - (−2,12.2) apply
    - (−2,0) apply and close
    - Right click Sketch 1

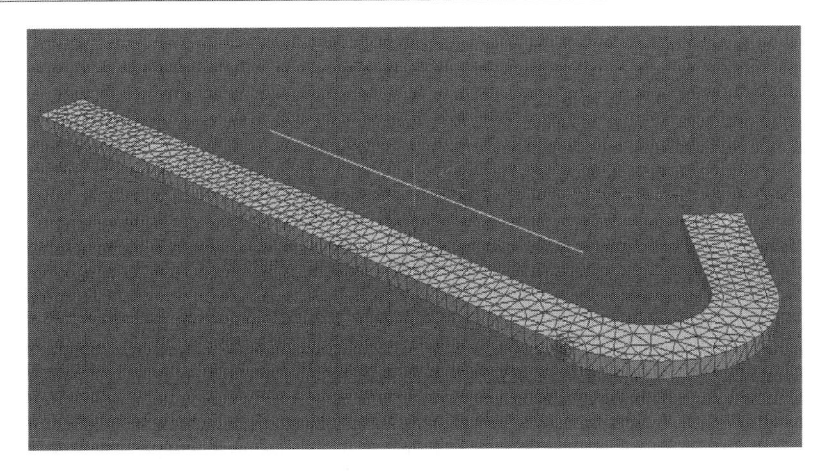

Figure A.4 A snapshot of a jet discharged into a channel bend.

- Rename as "Entrance"
- The updated geometry is shown in the following figure.

2 Straight Part of Exit Section (Figure A.6)

- New Entity/Basic/2D Sketch
- (0,0), apply
- (2.4,0) apply
- (2.4,−1) apply
- (0,−1) apply and close
- Right click Sketch 1

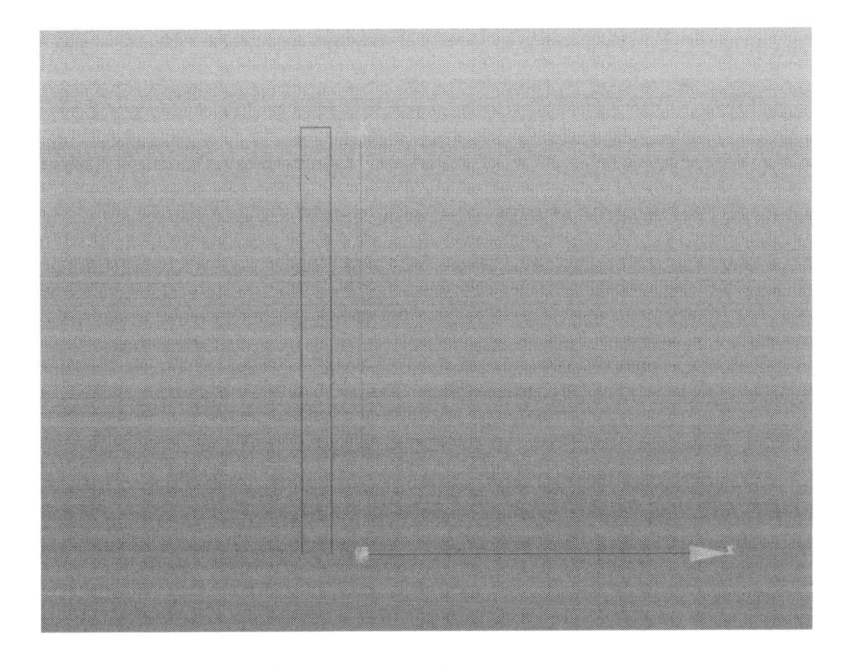

Figure A.5 A snapshot of the straight entrance section.

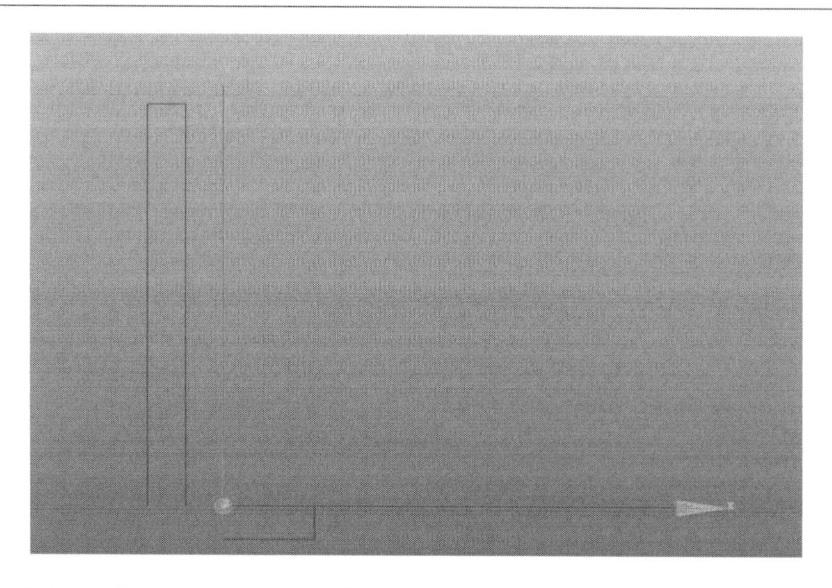

Figure A.6 A snapshot of part of the straight exit section.

- Rename as "Exit"
- The updated geometry is shown in the following figure.

3 Path Line of Straight Exit Section (Figure A.7)

- New Entity/Basic/2D Sketch
- (0,0), apply
- (1,0), apply and close
- Right click Sketch 1
- Rename as "PathLine"

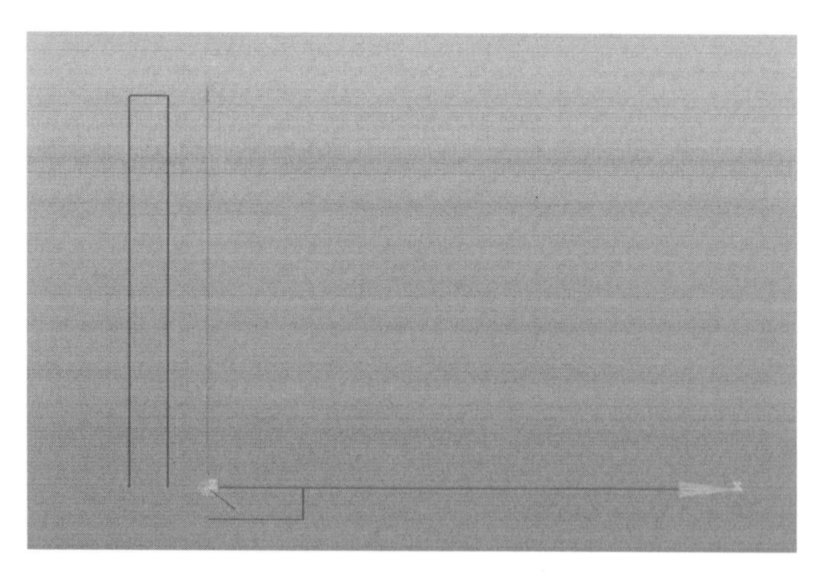

Figure A.7 A snapshot of path line of straight exit section.

- Operations/Transformation/Rotation
- Objects: Pathline
- Axis: OZ
- Angle: −45
- De-select "Create a copy"
- Apply and Close
- The updated geometry is shown in the following figure.

4 Final Exit Section of Straight Exit Section

- Operations/Transformation/Rotation
- Objects: Exit
- Axis: OZ
- Angle: 45
- De-select "Create a copy"
- Apply and Close
- Operations/Transformation/Modify Location
- Select the 3rd mode
- Objects: Exit
- Path object: PathLine
- Distance: 1
- De-select "Create a Copy"
- Apply and Close
- The updated geometry is shown in Figure A.8

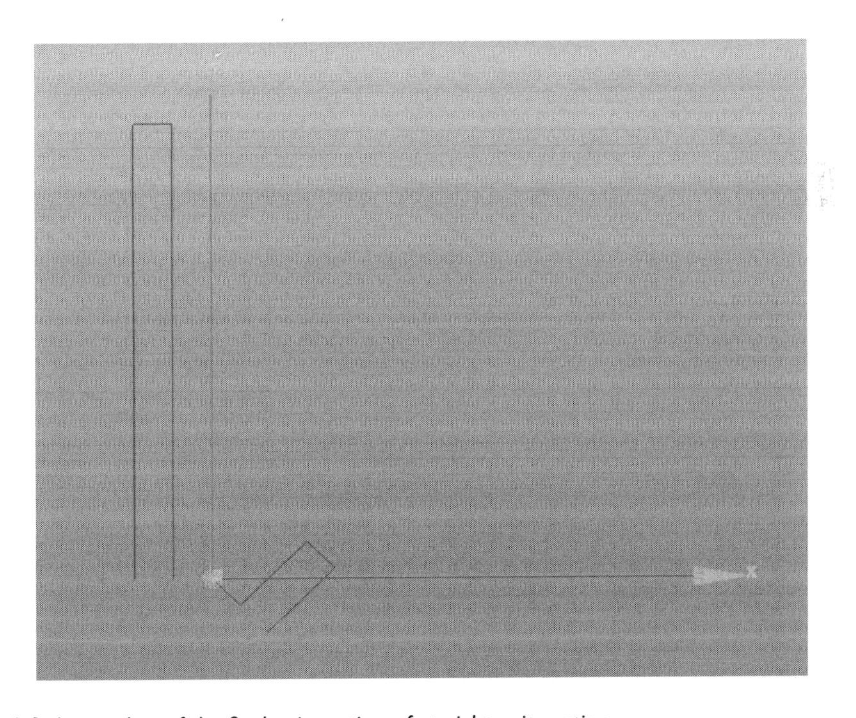

Figure A.8 A snapshot of the final exit section of straight exit section.

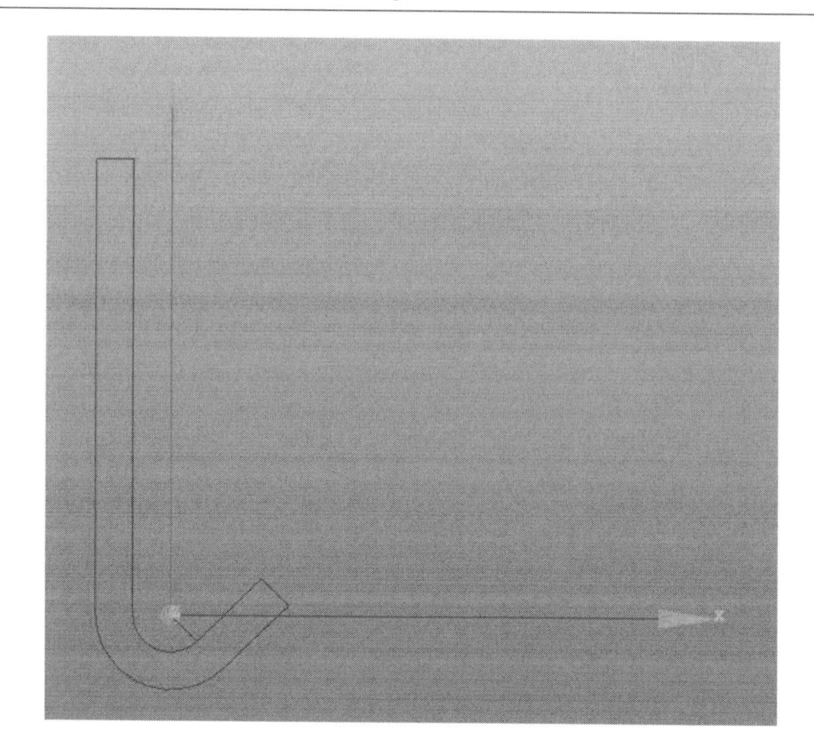

Figure A.9 A snapshot of the bend section.

5 Bend Section

- New Entity/Basic/Arc
- Select the 2nd mode
- Select the origin point (the start point of the PathLine) as Center Point
- Also set the Point Start, and Point End for the two arcs
- Apply and Close
- Rename as "InnerArc" and "OuterArc", respectively
- The updated geometry is shown in Figure A.9

6 Build Face of the Bend Section

- New Entity/Build/Face
- Name: Bottom
- Objects: Entrance; Exit; InnerArc; and OuterArc
- Apply and CloseOperations/Transformation/Rotation
- Objects: Pathline
- Axis: OZ
- Angle: −45
- De-select "Create a copy" Studyl
- Apply and Close
- The updated geometry is shown in Figure A.10

7 Extrusion of the Bend Section

- New Entity/Generation/Extrusion
- Name: Channel

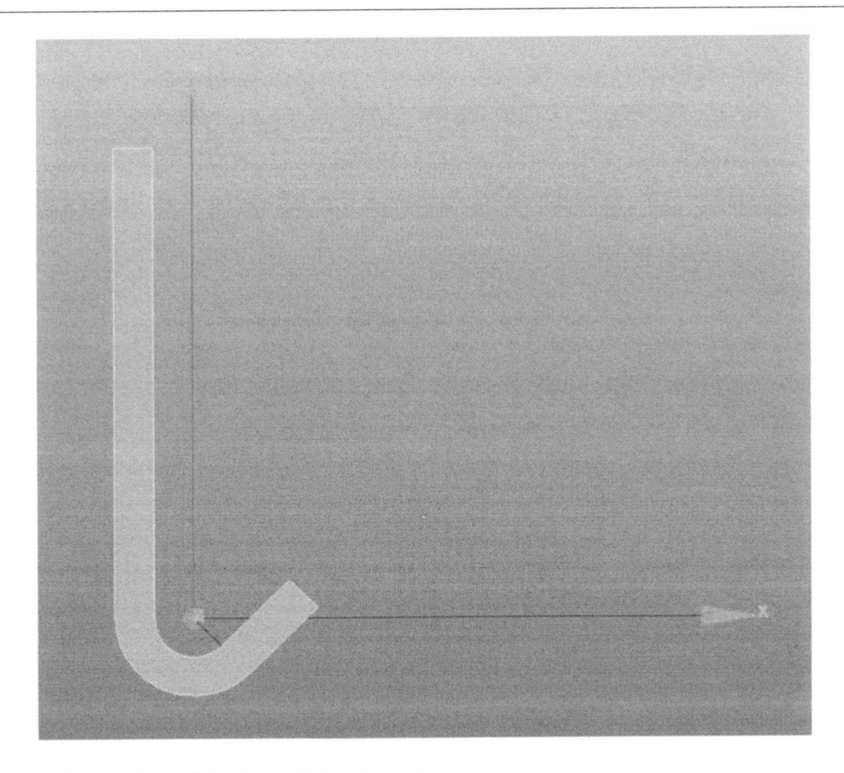

Figure A.10 A snapshot of the face of the channel.

- Base: Bottom
- Vector: OZ
- Height: 0.2
- Apply and Close
- The updated geometry is shown in Figure A.11

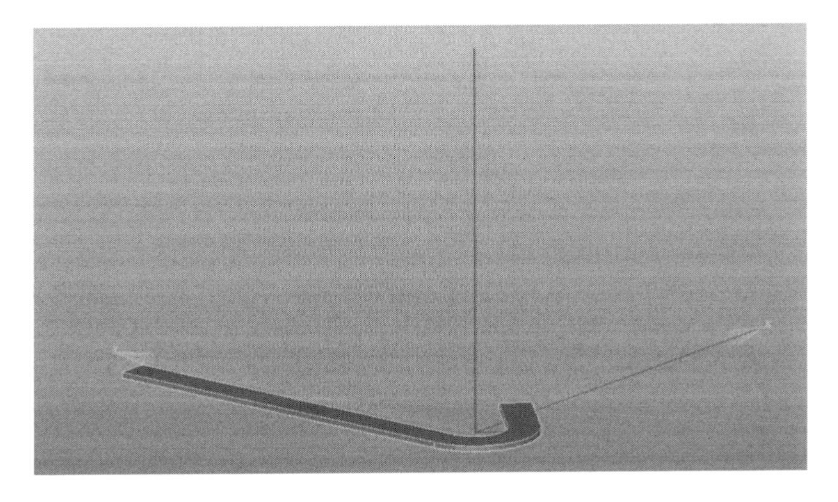

Figure A.11 A snapshot of the channel.

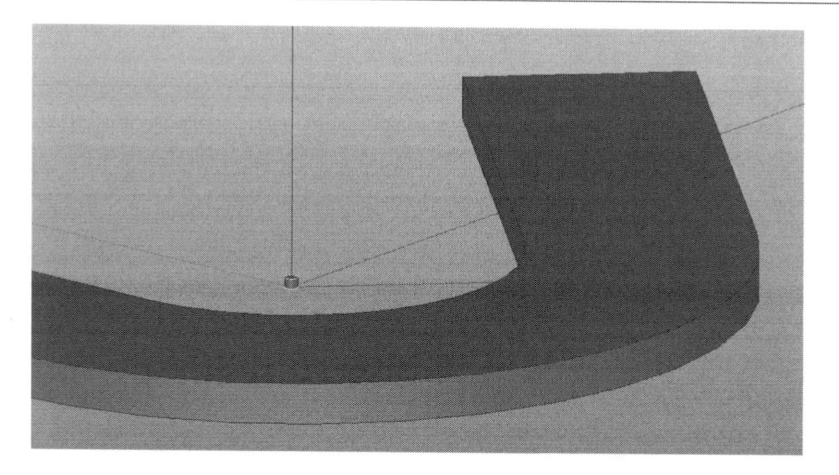

Figure A.12 A snapshot of the vertical nozzle.

8 Horizontal Nozzle

- New Entity/Primitives/Cylinder
- Name: Nozzle
- Radius: 0.03
- Height 0.05
- Apply and Close
- The updated geometry is shown in Figure A.12

9 Rotate Nozzle

- Operations/Transformation/Rotation
- Objects: Nozzle
- Axis: OY
- Angle: 90
- De-select "Create a copy"
- Apply and Close
- The updated geometry is shown in Figure A.13

10 Change Location of the Nozzle

- Operations/Transformation/Translation
- Objects: Nozzle
- Dx=−2
- Dy=0
- Dz=0.1
- De-select "Create a copy"
- Apply and Close

11 Create the Body of the Nozzle

- Operations/Boolean/Cut
- Name: ChannelWithNozzle
- Main Object: Channel
- Tool Objects: Nozzle
- The updated geometry is shown in Figure A.14

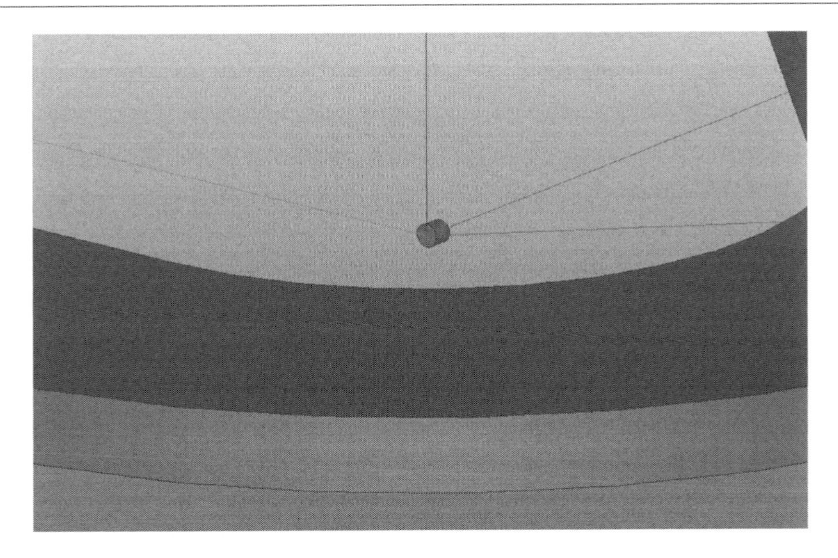

Figure A.13 A snapshot of the rotated nozzle.

12 Explode to Faces

- New Entity/Explode
- Main Object: ChannelWithNozzle
- Subshapes Type: Face
- Apply and Close

13 Determine the Groups and Faces

- The desired groups are determined and summarized in the following table; right click the faces, and click "show only" to clearly view the face, and fill them into the following table:

Group Name	*Main Shape*
NozzleInlet	Face_2
NozzleWall	Face_1
ChannelInlet	Face_9
ChannelOutlet	Face_12
ChannelWall:	Face_3; Face_4; Face_5; Face_7; Face_8; Face_10; Face_11;
ChannelUp	Face_6

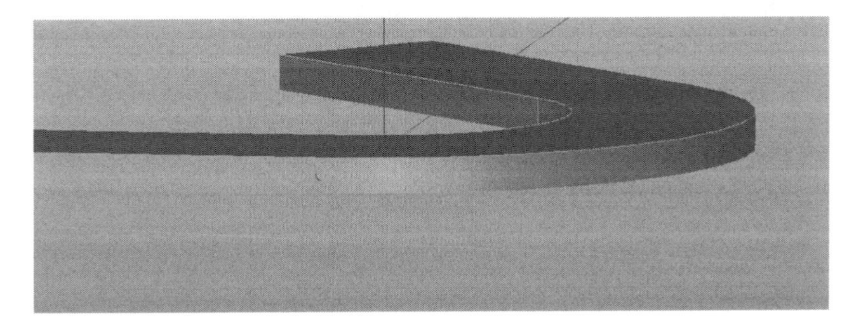

Figure A.14 A snapshot of the body of the nozzle.

14 Rename the Faces and Create Groups

- New Entity/Group/Create Group
- Select the 3rd mode
- Enter the values in the table above for the groups with multiple faces (ChannelWall)
- Name: ChannelWall
- Main Shape: ChannelWithNozzle
- Add the faces
- Apply and Close
- Rename the other Faces

15 Create Mesh

- Select the Mesh Module
- Select ChannelWithNozzle
- Mesh/Create Mesh
- Algorithm: Netgen 1D-2D-3D
- Hypothesis: NETGEN 3D Parameters
- Fineness: Very Coarse
- Ok
- Apply and Close
- Right click mesh1
- Compute
- The updated geometry is shown in Figure A.15

16 Create Group (Figure A.16)

- Mesh/Create Group from Geometry
- Element Type: Face
- Geometry: select all the renamed faces or group
- Apply and Close

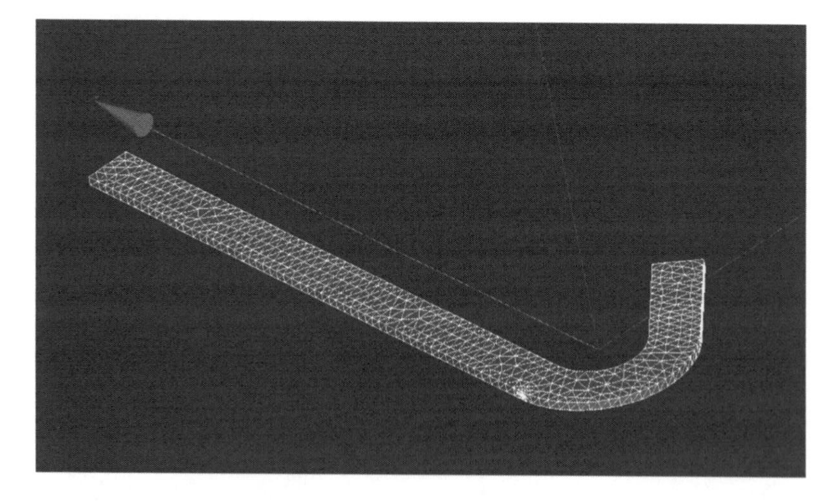

Figure A.15 A snapshot of the mesh.

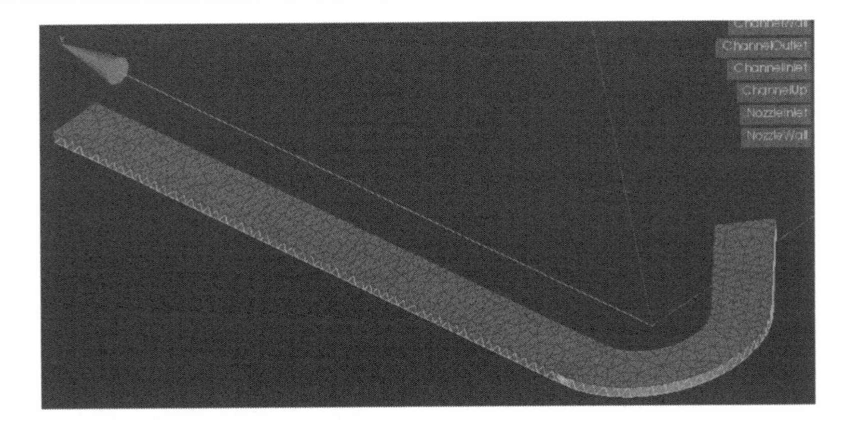

Figure A.16 A snapshot of the mesh with groups.

17 Export Mesh

- Right-click Mesh
- Export/UNV File
- Name as "MCBMesh1.unv"

18 Convert to OpenFOAM Format

- Delete the original polyMesh folder
- Copy the mesh file to the main folder of a case
- Type "ideasUnvToFoam MCBMesh1.unv"
- checkMesh

19 Check the Mesh in ParaView

- paraFoam
- Select all of the Mesh Parts
- De-select all of the Volume Fields
- Apply
- Surface with Edges
- The updated geometry is shown in Figure A.17

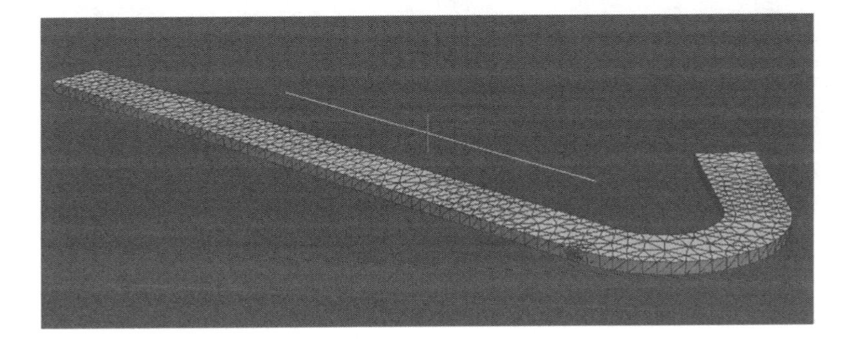

Figure A.17 A snapshot of the mesh shown in ParaView.

A.7 Mesh generation using the SnappyHexMesh utility

A.7.1 *Tutorial 3: Effluents discharged into a domain with obstacles*

This section provides a step-by-step tutorial on how to generate the mesh using the SnappyHexMesh utility for the simulation of effluents discharged into a domain with obstacles by modifying sample SnappyHexMesh configuration files. A snapshot of the final mesh is shown in Figure A.18.

1 Preparation

   1 Copy the basic files from a tutorial example.
   2 Create a folder for snappyHexMesh STL file: constant/triSurface.
   3 Copy the STL files into the new folder, the geometry of the STL files, which represent obstacles or walls, as illustrated in the following figures

2 Configure blockMeshDict to setup the computational domain

   1 Define the vertices

```
vertices
(
        (0 0 -11)//0
        (160 0 -11)//1
        (0 6 -11)//2
        (160 6 -11)//3
        (0 0 0)//4
        (160 0 0)//5
        (0 6 0)//6
        (160 6 0)//7
        (0 0 11)//8
```

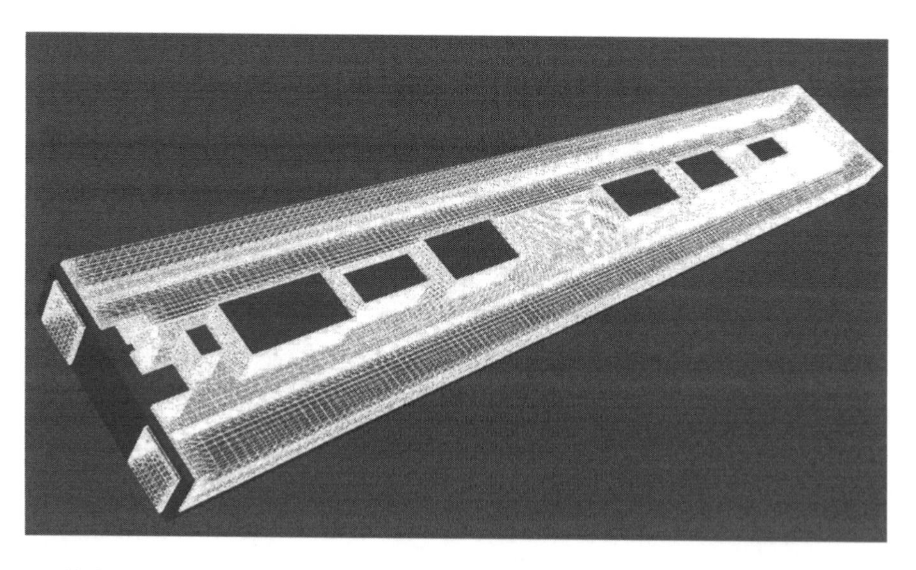

Figure A.18 Snapshot of the mesh for effluents discharged into a domain with obstacles.

Figure A.19 The geometry of the STL files.

```
        (160 0 11)//9
        (0 6 11)//10
        (160 6 11)//11
);
```

2 Define the blocks

```
blocks
(
    hex (0 1 3 2 4 5 7 6) (160 12 22) simpleGrading (1 1 1)//0
    hex (4 5 7 6 8 9 11 10) (160 12 22) simpleGrading (1 1 1)//0
);
```

3 Copy the STL files into the new folder, the geometry of the STL files, which represent obstacles or walls, as illustrated in the following figures (Figure A.18 and A.19)

4 Define the boundaries

```
boundary
(
    ob1//open boundary 1; should be inlet
    {
      type patch;
      faces
      (
          (4 6 10 8)
      );
    }
    ob2//open boundary 2; should be outlet
    {
      type patch;
      faces
      (
          (0 2 6 4)
```

```
        );
    }
    ob3//open boundary 3; should be outlet
    {
        type patch;
        faces
        (
        (5 7 11 9)
        );
    }
    ob4//open boundary 4; should be outlet
    {
        type patch;
        faces
        (
            (1 3 7 5)
        );
    }
    walls
    {
        type wall;
        faces
        (
            (0 1 5 4)
            (4 5 9 8)
            (0 1 3 2)
            (8 9 11 10)
            (2 3 7 6)
            (6 7 11 10)
        );
    }
);
```

3 Set mesh configurations for snappyHexMesh

1 Copy the files “surfaceFeaturesDict”, “snappyHexMeshDict”, and “meshQualityDict” from an example case (e.g., tutorials/mesh/snappyHexMesh/“flange”).
2 Modify surfaceFeaturesDict

- Replace the names of STL files

```
IO_WallA
{
        surfaces
        (
                "IO_WallA.stl"
        );
                // Identify a feature when angle between faces <
includedAngle
        includedAngle   150;
}
IO_WallB
{
```

```
        surfaces
        (
                "IO_WallB.stl"
        );
                // Identify a feature when angle between faces <
includedAngle
        includedAngle   150;
}
x0_block
{
        surfaces
        (
                "x0_block.stl"
        );
                // Identify a feature when angle between faces <
includedAngle
        includedAngle   150;
}
```

3 Modify snappyHexMeshDict

- Replace the names of stl files in the "geometry" section

```
geometry
{
     IO_WallA
     {
         type triSurfaceMesh;
         file "IO_WallA.stl";
     }
     IO_WallB
     {
         type triSurfaceMesh;
         file "IO_WallB.stl";
     }
     x0_block
     {
         type triSurfaceMesh;
         file "x0_block.stl";
     }
};
```

- Replace the names of STL files in the "features" section

```
  features
     (
         {
             file "IO_WallA.extendedFeatureEdgeMesh";
             level 0;
         }
         {
             file "IO_WallB.extendedFeatureEdgeMesh";
             level 0;
         }
```

```
        {
            file "x0_block.extendedFeatureEdgeMesh";
            level 0;
        }
);
```

- Replace the names of STL files in the "refinementSurfaces" section

```
refinementSurfaces
{
    IO_WallA
    {
        // Surface-wise min and max refinement level
        level (2 2);
    }
    IO_WallB
    {
        // Surface-wise min and max refinement level
        level (2 2);
    }
    x0_block
    {
        // Surface-wise min and max refinement level
        level (2 2);
    }
}
```

- Modify the locationInMesh: inside the domain but outside the objects

```
locationInMesh (10 1 6); // Inside point
```

- Replace the names of STL files in the "layers" section

```
layers
{
    "IO_WallA_.*"
    {
        nSurfaceLayers 1;
    }
    "IO_WallB_.*"
    {
        nSurfaceLayers 1;
    }
    "x0_block_.*"
    {
        nSurfaceLayers 1;
    }
}
```

- Keep the other settings default

4 Run the code for snappyHexMesh

1. blockMesh
2. surfaceFeatures
3. snappyHexMesh -overwrite
4. It shows "Finished meshing without any errors"

Index